UNDERGRADUATE TEXTS IN CONTEMPORARY PHYSICS

Series Editors
John P. Ertel
Robert C. Hilborn
David Peak
Thomas D. Rossing
Cindy Schwarz

Springer

New York
Berlin
Heidelberg
Hong Kong
London
Milan
Paris
Tokyo

UNDERGRADUATE TEXTS IN CONTEMPORARY PHYSICS

Cassidy, Holton, and Rutherford, Understanding Physics

Enns and McGuire, Computer Algebra Recipes: A Gourmet's Guide to the Mathematical Models of Science

Hassani, Mathematical Methods: For Students of Physics and Related Fields

Hassani, Mathematical Methods Using *Mathematica®:* For Students of Physics and Related Fields

Holbrow, Lloyd, and Amato, Modern Introductory Physics

Liboff, Primer for Point and Space Groups

Möller, Optics: Learning by Computing, with Examples Using Mathcad®

Roe, Probability and Statistics in Experimental Physics, Second Edition

Rossing and Chiaverina, Light Science: Physics and the Visual Arts

PRIMER FOR POINT AND SPACE GROUPS

Richard L. Liboff

With 39 Illustrations

 Springer

Richard L. Liboff
School of Electrical Engineering
Cornell University
Ithaca, NY 14853
richard@ee.cornell.edu

Series Editors

John Ertel
Department of Physics
United States Naval Academy
572 Holloway Road
Annapolis, MD 21402-5026
USA
jpe@nadn.navy.mil

Robert C. Hilborn
Department of Physics
Amherst College
Amherst, MA 01002
USA

David Peak
Department of Physics
Utah State University
Logan, UT 84322
USA

Thomas D. Rossing
Science Department
New Trier High School
Winnetka, IL 60093
USA

Cindy Schwarz
Department of Physics
Northern Illinois University
De Kalb, IL 60115
USA

Library of Congress Cataloging-in-Publication Data
Liboff, Richard L., 1931–
    Primer for Point and Space Groups / Richard Liboff
        p. cm. – (Undergraduate texts in contemporary physics)
    Includes bibliographical references and index.

    1. Group theory.   I. Title.   II. Series
    QC20.7.G76L53 2003
    512′.2—dc21   2003050662

ISBN 978-1-4419-2317-2

Printed in the United States of America.   (SBA)

9 8 7 6 5 4 3 2 1

Springer-Verlag is a part of *Springer Science+Business Media*

springeronline.com

For my granddaughter,
Scarlet Grace Raphaella

Preface

This text stems from a course I have taught a number of times, attended by students of material science, electrical engineering, physics, chemistry, physical chemistry and applied mathematics. It is intended as an introductory discourse to give the reader a first encounter with group theory. The work concentrates on point and space groups as these groups have the principal application in technology. Here is an outline of the salient features of the chapters.

In Chapter 1, basic notions and definitions are introduced including that of Abelian groups, cyclic groups, Sylow's theorems, Lagrange's subgroup theorem and the rearrangement theorem. In Chapter 2, the concepts of classes and direct products are discussed. Applications of point groups to the Platonic solids and non-regular dual polyhedra are described. In Chapter 3, matrix representation of operators are introduced leading to the notion of irreducible representations ('irreps'). The Great Orthogonality Theorem (GOT) is also introduced, followed by six important rules relating to dimensions of irreps. Schur's lemma and character tables are described. Applications to quantum mechanics are discussed in Chapter 4 including descriptions of the rotation groups in two and three dimensions, the symmetric group, Cayley's theorem and Young diagrams. The relation of degeneracy of a quantum state of a system to dimensions of irreps of the group of symmetries of the system are discussed, as well as the basis properties of related eigenfunctions. Schur's lemma is revisited in the construction of irreps of $SU(2)$. The relation of irreps of the rotation group to coupled angular momentum states is described. A section on degenerate perturbation theory is included.

A review is included in Chapter 5 of basic solid-state concepts including that of primitive vectors, the reciprocal lattice and the Brillouin zone. Cosets, invariant subgroups and the factor group come into play in this chapter and are applied to solid-state physics in which the real affine group, the Seitz operator and the space group of a lattice are defined. The 230 space groups and 32 crystallographic point groups are discussed. 'The group of $\mathbf{k}$,' 'the space group of $\mathbf{k}$' and 'the star of $\mathbf{k}$' are described as well. The chapter concludes with a discussion of double space groups, important for a description of spin-orbit coupling in crystals. Further applications to crystal physics occur in Chapter 6 with descriptions included of correlation diagrams, and electric and magnetic point groups. In this chapter, the phenomena of polarizability, the piezoelectric effect as well as 'black and white' and 'grey' groups come into play. A section on tensors in group theory is included with discussions of the 'general linear group' in all positive integer dimensions and the concept of irreducible tensors. The chapter concludes with a description of time reversal and space inversion in a crystal.

The final chapter is intended for the more mathematically inclined, and descriptions of elements of abstract algebra leading to the Galois group are included. A number of concepts such as invariant subgroups and factor groups reappear in this chapter (previously labeled, 'quotient groups' and 'normal subgroups'). Integral domains, fields, rings and congruences are defined and the notion of irreducible and symmetric polynomials are introduced. The concept of a solvable group comes into play in the description of the Galois group. A list of symbols is included for this final chapter.

A set of problems appears at the conclusion of each chapter preceded by a 'summary of topics' for the given chapter. A number of problems carry solutions. Thus, for example, a problem in Chapter 2 employs Euler's polyhedron formula to derive the property that there are only five regular polyhedra and a problem in Chapter 4 employs Young diagrams to derive the numbers of irreps of the permutation group of order 6. The dimensions of irreps of the $O(2)$ group and their basis functions are described in another problem in this chapter. A problem in Chapter 5 illustrates that the point group of a direct lattice applies as well to its reciprocal lattice. In this context, the text may prove useful for self-study.

A bibliography is included listing a number of related books in group theory. Character tables of the point groups are listed in Appendix A. Appendix B lists dimensions and notations related to the O_h and D_{4h} groups important to the double space groups discussed in Chapter 5. Figures are distributed throughout the text.

This work should stand as a springboard from which the student may pursue further study in any of the more advanced works cited in the bibliography.

A number of individuals have been of assistance in the writing of this text and I take this opportunity to express my deep appreciation to these kind individuals: Robert Connelly, Richard Freedberg, V. Veervari, Peter

Krusius, N. David Mermin, Rao Subramanya, Keith Dennis, Ali Lopez, Peter Lepage, Veit Elser, Clifford Earle and Kurt Gottfried. I am indebted to Mason Porter and Rajit Manohar for their careful reading of Chapter 7 and am particularly indebted to Robert Fay for sharing his deep knowledge of chemical applications of group theory with me.

תושלב"ע   R.L.L., Ithaca, 2002

Contents

1
Groups and Subgroups

1.1 Definitions and Basics

Group theory is the study of symmetries, with a very wide domain of applications. In this work we are interested in laying the groundwork for application to solid-state physics and other areas of technology. The basic notions developed in the first five chapters of the present work are universal and apply to many areas of application.

A group is a collection of distinct elements, $A, B, \ldots$, with a binary operation (often called the 'product') which obey the following rules:

(a) The product of any two elements of the group is a member of the group. The square of any element of the group is a member of the group. Let G denote the group and let A, B be any two elements of G. Then if $AB = C, C \in G$. Similarly, if $AA = D, D \in G$. This property is called *closure*.

(b) One element of the group, E, is called the *identity operator*. The operator E commutes with every element of G and leaves it unchanged. If A is any element of G, then

$$EA = AE = A \qquad (1.1a)$$

(c) Members of the group obey the *associative law*:

$$A(BC) = (AB)C \qquad (1.1b)$$

(d) Every member of the group has a *reciprocal* (or *inverse*). The group element B^{-1} is the inverse of A iff

$$AB^{-1} = B^{-1}A = E \qquad (1.1c)$$

Here are four common definitions in group theory:

Abelian Group[1]

A group is Abelian iff all members of the group commute. Thus, if A, B are any two elements of an Abelian group, then

$$AB - BA \equiv [A, B] = 0 \qquad (1.2)$$

where $[A, B]$ is called the *commutator* of A and B.

Order of a Group

The order of a group is the number of elements in the group. This parameter carries the symbol "h." An example of an Abelian group with infinitely many members ($h = \infty$) is that of all integers under addition:

$$\begin{aligned} &\text{Identity: } E = 0 \ (n + 0 = n) \\ &\text{Inverse of } n = -n \ [n + (-n) = 0 = E] \\ &\text{Abelian property: } 1 + 3 = 3 + 1. \end{aligned} \qquad (1.3)$$

The set $\{1, -1\}$ under multiplication, comprises a group of order 2 ($h = 2$). Each element is its own inverse and the identity is 1. Note that in the first of these examples, the binary operation in rule (a) above, is addition, whereas in the second example it is multiplication.

Point Group

The operation of an element of a point group on an object maps the object onto itself and leaves one point fixed. Consider, for example, rotations in the plane of an equilateral triangle about its point of symmetry (Fig. 1.1). As noted in this figure, these symmetries comprise a set of three operations, labeled (C_3, C_3^2, E). This set of operations comprises a group of order 3 ($h = 3$). The rotation operation C_3, rotates the triangle through $2\pi/3$. More generally, the operation C_n rotates a figure in the plane, about its center of symmetry, through $2\pi/n$, where n is a positive integer. It will be noted that C_n denotes a rotation through $2\pi/n$ and is also the label of the related point group. Thus, for example, the point group of symmetries of the equilateral triangle in the plane is labeled C_3.

---

[1]Named for N. H. Abel (1802–1829).

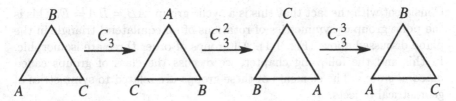

Figure 1.1. Elements of the C_3 point group. Rotations of the equilateral triangle.

The Cyclic Group

A cyclic group contains the elements: $(A_n, A_n^2, A_n^3, \ldots, A_n^{(n-1)}, E)$, is of order n, and has the property, $A_n^n = E$. The set of rotations in the plane of the equilateral triangle, described above, is an example of a cyclic group of order three. All cyclic groups are Abelian. For example,

$$A_n^2 A_n^5 = A_n^5 A_n^2 = A_n^7 \tag{1.4a}$$

$$A_n^{n+k} = A_n^k, \ \ 1 \le k \le n \tag{1.4b}$$

The set of rotations in the plane of a regular n-sided polygon about its point of symmetry, which reproduces the polygon, is the cyclic group $(C_n, C_n^2, C_n^3, \ldots, C_n^{(n-1)}, E)$. This group is labeled C_n.

1.2   Group Table

Having defined a group, the next fundamental concept in group theory is that of the *group* (or *multiplication*) table. The multiplication table of a group of order h is an $h \times h$ square array consisting of all products of the group. For example, the multiplication table of the group of order two, (A, E), which we label C_2, has four entries.

C_2	E	A
E	EE	AE
A	AE	AA

or, equivalently,

C_2	E	A
E	E	A
A	A	E

(1.5a)

Note that if $AA = A$ then $A = E$ and the group reduces to the group of order one. The group of order one contains only the identity operator and is labeled C_1.

The multiplication table for the group of order three, labeled C_3, is given by

C_3	E	A	B
E	E	A	B
A	A	B	E
B	B	E	A

(1.5b)

Consistent with the fact that this is a cyclic group, $AB = BA = E$. This is the point group of symmetries of rotations of an equilateral triangle in the plane discussed above (Fig. 1.1). All groups of order three are isomorphic. In this and the following chapter, we discuss the class of groups called *dihedral groups*. The elements of these groups are related to symmetries of geometrical objects.

1.3   Rearrangement Theorem

The rearrangement theorem states that each row and each column of a group multiplication table consists of all the group elements once and only once. To prove this theorem we consider the partial group table shown below, for which A_1 appears twice in the first column,

$$
\begin{array}{c|ccccccc}
G_n & E & A_1 & A_2 & A_3 & . & . & . \\
\hline
E & E & A_1 & A_2 & A_1 & . & . & . \\
A_1 & . & . & . & . & . & . & . \\
A_2 & . & . & . & . & . & . & .
\end{array}
$$
(1.6)

According to the first row of this listing, $EA_1 = A_1 = EA_3 = A_3$, which implies that $A_1 = A_3$ which is a contradiction, as all elements of a group are distinct.

The rearrangement theorem is very useful in construction of the group table for a specific group. For example in any of the cases (1.5), as well as any other finite group, if an element of a row or column of the related group table is missing, this missing element is determined by the rearrangement theorem.

We may use the rearrangement theorem to prove that the point group C_3, whose group table is given by (1.5b), is unique. In the remaining choice of construction of this group, one sets diagonal elements equal to E. There results

$$
\begin{array}{c|ccc}
C_3 & E & A & B \\
\hline
E & E & A & B \\
A & A & E & \\
B & B & & E
\end{array}
$$

The rearrangement theorem indicates that the missing entry in the last column of the above array is A so that $AB = A$. The missing entry in the last row is likewise A so that $BA = A$. The repetition of A in both the second row and second column violates this theorem. One may conclude that the C_3 group is unambiguously given by (1.5b).

The i and σ Operations

We have encountered the E and C_n operations. Two additional operations are described as follows. In the inversion operation, labeled "i", an object is inverted through its center of symmetry and is described by the transformation of coordinates, $(x, y, z) \rightarrow (-x, -y, -z)$. In the reflection operation, labeled σ, an object is reflected through a plane of symmetry. We apply these operations to examples of groups of order two with related geometrical configurations. These examples are shown in Fig. 1.2. In case (a), that of the rhomboid, inversion (i) through the symmetry point (the point of intersection of the two diagonals), reproduces the figure. In case (b), that of the isosceles triangle, reflection (σ) through the bisecting line returns the original figure. In case (c), that of the one-dimensional linear symmetric molecule, rotation (C_2) about the center, B, through π, returns the original figure. Note for that this latter case, $\sigma = C_2$. This case of the linear symmetric molecule may also be pictured as two points, A, at the ends of the diameter of a circle of radius AB. Rotation through π returns the original figure. The group tables for these second-order groups are shown below,

C_i	E	i
E	E	i
i	i	E

C_σ	E	σ
E	E	σ
σ	σ	E

C_2	E	C_2
E	E	C_2
C_2	C_2	E

C_{xp}	$\exp(i0)$	$\exp(i\pi)$
$\exp(i0)$	$\exp(i0)$	$\exp(i\pi)$
$\exp(i\pi)$	$\exp(i\pi)$	$\exp(i0)$

$$(1.7)$$

The C_{xp} group is defined under standard multiplication. Note that in the C_{xp} group, $i = \sqrt{-1}$, and should not be confused with the group theoretical definition of i, the inversion operator, such as occurs in the C_i group. The identity element of C_{xp} is $\exp(i0) = 1 = \exp(i2\pi)$. In all four cases, the nonidentity element of the group is its own inverse. The σ plane of reflection in the C_S group corresponds to the symmetries of the object at hand.

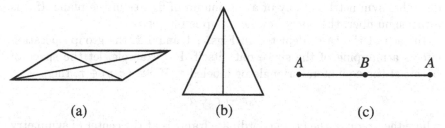

(a)   (b)   (c)

Figure 1.2. Examples of geometrical figures in the plane with symmetries described by second-order groups: (a) the rhomboid, which obeys i symmetry; (b) the isosceles triangle, which obeys σ symmetry; (c) the idealized one-dimensional linear symmetric molecule, which obeys C_2 symmetry.

Homomorphism and Isomorphism

It is evident that all four groups whose group tables are shown in (1.7) are equivalent, namely, there is a one-to-one correspondence between elements of any two groups in this set. For example, to demonstrate equivalence of the C_i and C_σ groups, one sets $i \to \sigma$. In this context, one introduces the notions of homomorphic and isomorphic mappings. Let f be a mapping of the elements $g \in G$ onto the elements $g' \in G'$. That is, $f(g) = g'$. The mapping f is *homomorphic* iff

$$f(gg') = f(g)f(g') = gg' \tag{1.8a}$$

In a homomorphic mapping, $f(g)$, of the elements of G onto the elements $g' \in G'$, the set of those elements of G which map onto the identity of G' is called the *kernel* of F. That is, the set of elements, $g \in G$, for which

$$f(g) = E \in G' \tag{1.8b}$$

is called the kernel of G. Two groups G and G' related by a homomorphic mapping are said to be *homomorphic*. This relation is written $G \sim G'$. A homomorphic mapping may be a many-to-one transformation. If the mapping is one-to-one and satisfies (1.8a) it is an *isomorphic* mapping. (Any isomorphism is a homomorphism.) With relabeling, the multiplication tables of two isomorphic groups are the same. Thus, all four groups whose group tables are shown in (1.7) are isomorphic. A theorem of group theory (to be established) states that there is only one group whose order is a prime number. That is, all groups of a specific prime order are isomorphic. We have seen examples of this theorem for the cases $h = 2$ and $h = 3$.

The preceding elementary description of the C_2 group demonstrates, in part, the deep significance of group theory. It is evident that this point group describes symmetries of any of a continuum of figures in the plane. Furthermore, as there is only one C_2 group, its group table is independent of the specific "multiplication" brought into play in definition of the corresponding application. Similarly, any of the cyclic groups, C_n ($n \geq 2$), describes symmetries of any of a continuum of figures in the plane. If n is a prime number, the related cyclic group is unique.

In each of the cases depicted in Figs. 1.1 and 1.2, the group operation effects a mapping of the space onto itself. For example, let the space of points of the equilateral triangle be labeled τ. If $r \equiv (x, y) \in \tau$, then,

$$C_3 r \to r' \in \tau$$

where the origin of the (x, y) coordinate frame is at the center of symmetry of the triangle. We label the set of points on which a group operates the *domain* of that operator. In the preceding example, τ is the domain of C_3.

Note that for the one-dimensional molecule (Fig. 1.2(c)), $C_2(x, y) = i(x, y) = (-x, -y)$. In three dimensions, if the z axis is taken as the C_2 axis, then $C_2(x, y, z) = (-x, -y, z) \neq i(x, y, z)$.

The Linear Molecule

There are two types of linear molecules: (a) Those which consist of two equivalent halves (e.g., ClCl, OCO, NCCN,...). (b) Those with dissimilar halves (e.g., HCl, HNO,...). A linear molecule has a symmetry axis common to all nuclei.

For case (a) there are three sets of symmetries:

(i) Any perpendicular bisector axis of the symmetry axis is a C_2 axis. There are an infinite number of such axes.

(ii) Any plane which includes the symmetry axis is a symmetry plane. There is an infinite number of such planes.

(iii) A plane of symmetry perpendicular to the symmetry axis.

The corresponding group is called the $D_{\infty h}$ group. For case (b), only the symmetries (ii) maintain. The corresponding group is called the $C_{\infty v}$ group.

The $C_{\infty v}$ group contains the elements $(E, 2C_\phi, \sigma_v)$, where $2C_\phi$ denotes rotations through $\pm 2\phi$ about the symmetry axis and σ_v is a reflection in a symmetry plane of the molecule. The $D_{\infty h}$ group contains the elements $[E, 2C_\phi, i, 2iC_\phi, \sigma_v, C_2]$. Examples of objects whose symmetries are described by these groups are shown in Fig. 1.3. Example (a) is described by the $D_{\infty h}$ group and (b) by the $C_{\infty v}$ group. In either case, the figures are objects of rotation.

Generating Elements

A minimum number of elements of a point group which, through products of only these operators, produce the whole group are called *generators* of the group. For example, the operators C_3 and σ_c produce the C_{3v} group [see (1.10)]:

$$C_{3v} = \{E = \sigma_c^2, C_3, C_3^2 = C_3 C_3, \sigma_c C_3 = \sigma_a, \sigma_c C_3^2 = \sigma_b\}.$$

1.4   Building Groups. Subgroups

A uniform homogeneous prism has three principal axes. Two of these are normal to the third which lies along the extended direction of the prism. We refer to this third axis as the principal axis. The σ_v (v for vertical) operator reflects the prism through a plane which includes the principal axis. The σ_h (h for horizontal) operator reflects the prism through a plane which is normal to the principal axis. For the present discussion we concentrate on the σ_v operation. The C_2 group is relevant to the rotational symmetries of a rhombus or a rectangle in the plane (Fig. 1.4). It is evident that these figures have two lines of reflection symmetry, which meet at the center of the rhombus. We label these the x and y axes, respectively. Consider a

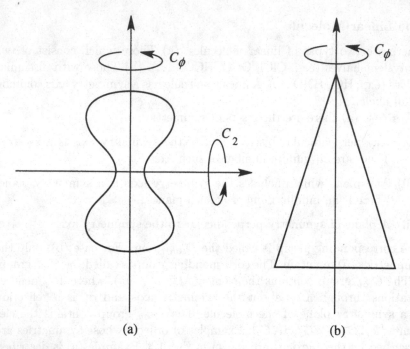

(a)   (b)

Figure 1.3. Each diagram represents a figure of rotation. Symmetries of the object are described by (a) the $D_{\infty h}$ group and those of the object and (b) by the $C_{\infty v}$ group.

uniform homogeneous prism whose cross section is, say, a rhombus. The σ_v^x operation reflects the prism through the plane $y = 0$, and the σ_v^y operation reflects the prism through the plane $x = 0$. Combining these operations with those of C_2 gives the fourth-order group, C_{2v}, whose group table is given by the following:

$$
\begin{array}{c|cccc}
C_{2v} & E & C_2 & \sigma_v^x & \sigma_v^y \\
\hline
E & E & C_2 & \sigma_v^x & \sigma_v^y \\
C_2 & C_2 & E & \sigma_v^y & \sigma_v^x \\
\sigma_v^x & \sigma_v^x & \sigma_v^y & E & C_2 \\
\sigma_v^y & \sigma_v^y & \sigma_v^x & C_2 & E
\end{array}
\tag{1.9}
$$

In addition to the rearrangement theorem assisting in the construction of this table, one also notes the following two generally valid relations:

$$\sigma^2 = E, \quad C_2^2 = E.$$

As is evident from the group table (1.9), the C_2 group is embedded in the C_{2v} group. Elements of a group, G, which themselves comprise a group are called a *subgroup* of G. A necessary condition for a set of elements of a group to be a subgroup is that the set includes the identity operator. Thus, for example, the group C_2 is a subgroup of C_{2v}. Note that the remaining

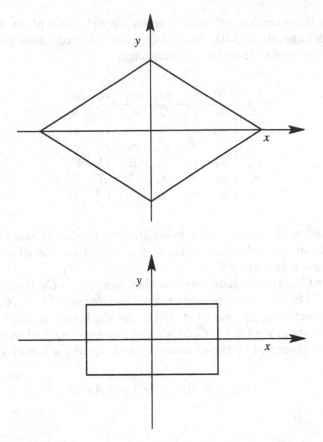

Figure 1.4. Two uniform homogeneous prisms whose respective cross sections (in the $z = 0$ plane) are: (i) the rhombus and (ii) the rectangle. In either case (neglecting reflections in the $z = 0$ plane) symmetries of these prisms are described by the C_{2v} group.

elements $(\sigma_v^x, \sigma_v^y, E)$ do not comprise a subgroup as $\sigma_v^x \sigma_v^y = C_2$ which is not an element of the proposed subgroup.

The C_{2v} group is interesting, because its order, 4, is not a prime number. With the stated prime-number theorem, one might expect another group of fourth order to exist which is not isomorphic to C_{2v}. This is the Abelian cyclic group, C_4, the point rotational symmetries in the plane, of the square.

The C_{3v} Group

The C_3 group table is given by (1.5b) and describes the point rotational symmetries of the equilateral triangle in the plane. With reference to Fig. 1.1, we note that the equilateral triangle has three lines of reflection symmetry, which individually are bisectors of respective vertex angles. Again, consider a uniform homogeneous prism whose cross section is now an equilateral triangle. Let the three planes of reflection symmetry be labeled

a, b, c. The σ_a operation reflects the prism through the a plane, etc. Combining these operations with those of C_3 gives the sixth-order group, C_{3v}, whose group table is given by the following:

C_{3v}	E	C_3	C_3^2	σ_a	σ_b	σ_c
E	E	C_3	C_3^2	σ_a	σ_b	σ_c
C_3	C_3	C_3^2	E	σ_c	σ_a	σ_b
C_3^2	C_3^2	E	C_3	σ_b	σ_c	σ_a
σ_a	σ_a	σ_b	σ_c	E	C_3	C_3^2
σ_b	σ_b	σ_c	σ_a	C_3^2	E	C_3
σ_c	σ_c	σ_a	σ_b	C_3	C_3^2	E

$$(1.10)$$

In constructing this group table, in addition to previously stated rules, it was noted, for example, that $C_3^2 C_3^2 = C_3$. Note that the Abelian cyclic group C_3 is a subgroup of C_{3v}.

With the C_{3v} group at hand one may show that $C_{3v} \sim C_2$. Recall that the group $C_2 = \{E, A\}$. In this homomorphism the elements $(E, C_3, C_3^{-1}) \to E$ and the elements $(\sigma_a, \sigma_b, \sigma_c) \to A$. Note that the kernel of this homomorphism is the subgroup (E, C_3, C_3^{-1}). Calling the mapping f, then with (1.8) one has, for example (for the elements C_3 and σ_a, of the two given sets),

$$f(C_3 \sigma_a) = f(C_3)f(\sigma_a) = EA = A$$

Subgroup Divisor Theorem (Lagrange's Theorem)

The group C_{3v} has two subgroups, E, for which $h = 1$, and the group C_3, for which $h = 3$. The fact that the orders of these subgroups both divide the order, 6, of C_{3v}, is part of a general theorem which states that the orders of subgroups of a group, individually divide the order, h, of the group. That is, $h/g = k$, where g is the order of a subgroup and k is a positive integer. Let us prove this. Consider a given group G with a subgroup, S, of order g, with the elements $A_1, A_2, \ldots A_g$. If $B \in G$ but $B \notin S$, then none of the products, $BA_1, BA_2, \ldots, BA_g$, are in S. Namely, suppose $BA_1 = A_2$. Let A_3 be the inverse of A_1. Then, $B = A_2 A_3$ which contradicts our assumption that $B \notin S$. It follows that all the products BA_i, as well as all the g elements of S are elements of G. Thus, there are at least $2g$ elements of $G : 2g < h$. Now choose an element $C \in G, C \notin S$ to obtain another g element in G so that $3g < h$. Continue this process until there are no more such sets. If k steps are taken to reach this state, then $h = kg$, where k is a positive integer, so that $h/g = k$, which was to be proved. The converse of this theorem, that if n divides h, then a subgroup of order n exists, is not in general valid.

Another sixth-order group not isomorphic to C_{3v} has the following group table:

$$
\begin{array}{c|cccccc}
G_6 & E & A & B & C & D & F \\
\hline
E & E & A & B & C & D & F \\
A & A & E & D & F & B & C \\
B & B & F & E & D & C & A \\
C & C & D & F & E & A & B \\
D & D & C & A & B & F & E \\
F & F & B & C & A & E & D \\
\end{array}
\tag{1.11}
$$

This group has the following subgroups: $[h = 1](E), [h = 2](E, A);$ $(E, B); (E, C), [h = 3](E, D, F)$, whose orders $1, 2, 3$ all are divisors of 6, the order of G_6. A subgroup that itself is not the whole group is a *proper subgroup*. A partial list of group operations, including those cited above, is given in Table 1.1.

Table 1.1

Group Operations

E	The identity operation.
i	The inversion operation.
C_n	Rotation through $2\pi/n$ about a symmetry axis. Sometimes called *the proper rotation operation*. By convention, rotation is taken in a right-handed sense.
σ	Reflection in a symmetry plane.
σ_h	Reflection in a plane that bisects the principal axis.
σ_v	Reflection in a plane that includes the principal axis.
σ_d	Reflection in a diagonal plane. This plane includes the principle axis and bisects a pair of two-fold axes perpendicular to the principal axis.
S_n	*The improper rotation operator*. This operation includes a proper rotation through $2\pi/n$ followed by a reflection in a plane perpendicular to the rotation axis: $S_n = \sigma_h C_n = C_n \sigma_h$. As $C_n^n = E$, and for n even, $\sigma_h^n = E$, it follows that $S_n^n = E$. In addition $S_2 = i$, the inversion operator. Similarly, for m even, $S_n^m = C_n^m$. For n odd, $S_n^n = \sigma_h$ and $S_1 = \sigma_h$ (see Fig. 1.5).

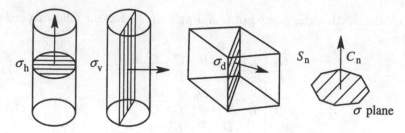

Figure 1.5. Four basic point-group reflections.

Let us show that S_2 is equivalent to the i operation. Let the z axis denote the axis of rotation so that C_2 operates in the (x, y) plane. Then

$$S_2(x, y, z) = \sigma C_2(x, y, z) \rightarrow \sigma(-x, -y, z) \rightarrow (-x, -y - z) = i(x, y, z)$$

Groups of Prime Order

We wish to show that a group of order $h, G(h)$, is unique when h is a prime number. Equivalently, all such groups are isomorphic.

Proof: Let $X \in G(h)$. Then there exists an integer n, such that $X^n = E$, as, $X \in G(h), X^2 \in G(h)$, etc. There are three possibilities: $n = 1, n = h$ or $n < h$. In the first two cases, n divides h. If $n < h$, then $\{X, X^2, \ldots, X^h = E\}$ is a subgroup of $G(h)$ and n divides h. If h is a prime number, then $n = 1$ or h. For the non-trivial case, the group so generated, $\{X, X^2, \ldots, X^h = E\}$, is unique and is cyclic.

Corollary. If $G(h)$ has a proper subgroup, then h is not a prime number. Let the subgroup be of order $n < h$. By Lagrange's theorem, n divides h, so that h is not a prime number.

Sylow's Theorems

Definition. Let p be a prime number. Then a group of order p^α for $\alpha \geq 1$, and integer, is a *p-group*. Subgroups of a group which are p-groups are called *p-subgroups*.

I. Consider a group G of order $p^n m$ where $n \geq 1$ and p does not divide m. Then G contains a subgroup of order p^r for each r where $1 \leq r \leq n$ (*Sylow p-subgroup*).

II. Let P_1 and P_2 be Sylow p-subgroups of a group G. Then P_1 and P_2 are conjugate subgroups of G (Chapter 2).

III. If p divides $h(G)$, then the number of Sylow p-subgroups of G is congruent to 1 (mod p) and divides h (that is, $n = 1 + kp$, for some integer k. See Chapter 7).

Chiral Property

A concept directly related to the improper S_n rotation operator is that of the chiral (Greek, hand) property of an object. An object is chiral if it is not congruent with any of its mirror images. Equivalently, one may say that an object is chiral iff it has no improper (S_n) rotations. Thus, an object with any number of C_n axes of symmetry but no S_n axis is chiral. Such is the case for the H_2O_2 (hydrogen peroxide) molecule (which has only C_2 symmetry). An object with a plane of mirror symmetry is not chiral. (The object is congruent with its image through that plane.) The human hand and a screw are chiral. A nail is achiral, as a nail has an infinite number of σ planes (in the axis of the nail), and $S_1 = \sigma$. In chemistry, a chiral molecule and its mirror image are called enantiomers (or, optically active isomers). When plane polarized light is passed through a medium of one type of enantiomer of a given molecule, the plane of polarization is rotated. The two possible directions of rotation carry the respective symbols, D (for dextro) and L (for levo). A medium composed equally of the two enantiomers of a given molecule is optically inactive. All naturally occurring amino acids (building blocks of proteins) except the simplest amino acid, glycine (which is achiral), are L-type. Any carbon atom with four distinct atomic groups lacks a plane of symmetry and is chiral. For example, glyceraldehyde, whose chemical diagram is shown in Fig. 1.6, is chiral. In real space, the top and bottom C-C bonds of the molecule drop beneath the page. The central C atom of the molecule is bonded to the four different atomic groups: CHO, H, OH and CH_2OH, and thus is chiral. Note that reflection of the molecule through the plane M gives the other enantiomer.

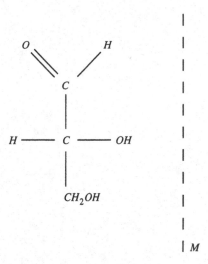

Figure 1.6. The chiral molecule, glyceraldehyde. The image enantiomer in the mirror M is incongruent with the molecule.

Summary of Topics for Chapter 1

1. Definition of groups: closure; existence of identity; associative law; existence of inverse.

2. Abelian groups.

3. Point groups.

4. Cyclic groups.

5. Rearrangement theorem.

6. Isomorphism and homomorphism.

7. Groups of the linear molecule.

8. Subgroups.

9. Lagrange's subgroup divisor theorem.

10. Generators of a group.

11. Prime-order group theorem.

12. List of group operations (Table 1.1).

13. Sylows' theorems.

14. Chirality.

Problems

1.1 What are the subgroups of the cyclic group, C_n, for (a) n even and (b) n odd?

1.2 Every cyclic group is Abelian. Is the converse of this statement true?

1.3 Sketch four plane figures whose symmetries are described by C_3.

1.4 Can two subgroups of a given group contain a common element? Justify your answer.

1.5 If $A, B \in G$, show that $(AB)^{-1} = B^{-1}A^{-1}$.
Answer

$$(AB)^{-1}AB = E$$
$$(AB)^{-1}A = B^{-1}$$
$$(AB)^{-1} = B^{-1}A^{-1}.$$

1.6 (a) What are the symmetry operations for the square prism? Accompany your answer with a figure. (b) Write down the group table for this group (C_{4v}). (c) Identify the subgroup of this group which is of largest order. (d) In what manner does the C_{4v} group change if two parallel edges of the square cross section of the prism are infinitesimally shortened?

1.7 Show, by construction of the related group table, in a similar manner as was done for C_3 in the text, that the fifth-order group is unique.

1.8 Show that, if a group of order h has a proper subgroup, h is not a prime number.

1.9 The top line of a character table of a group G appears as

G	E	A	B	C	D	F	G	H

where all 8 elements of G are shown. What type of group is G? Explain your answer. (b) What are the orders of the subgroups of this group?

1.10 (a) Show explicitly that a cyclic group of order 18 has proper subgroups with respective orders, 2 and 9. (b) Is this problem an application of one of Sylows' theorems? Explain your answer.

1.11 A given molecule has neither a plane of symmetry nor a center of symmetry. Is the molecule chiral?
Answer
We are told that the molecule has neither S_1 nor S_2 symmetries. If we are assured that the molecule has no higher order S_n symmetries, one may conclude that the molecule is chiral. If it has a higher order

S_n symmetry, then it is achiral (i.e., it is congruent with a mirror image of itself).

1.12 Is an object described by the C_1 group chiral or achiral?
Answer
An object described by the C_1 group has no S_n (nor any other) symmetry. Therefore it is chiral.

1.13 What is the symmetry group of a finite circular cone in three dimensions with a spherical cap of radius equal to that of the cone?

1.14 What are the generators of the group G_6 given by (1.11)?

1.15 Is the set of integers a group under multiplication?
Answer
No. As the inverse term $1/q$ is not an element of the group, where q is an integer, the set does not include inverses of elements. On the other hand, the set of rational numbers is a group under multiplication as the number p/q has the inverse q/p, which is an element of the group. The order of this group is infinite.

1.16 (a) Consider a cyclic group of order h. The group table forms an $h \times h$ matrix. State a property of this matrix. (b) Repeat part (a) for an Abelian group, G, with the properties $A_i^2 = E$ for all $A_i \in G$ and $h(G)$ is even.

1.17 What are the symmetry groups which describe, respectively: (a) The rhomboid? [A parallelogram with equal opposite sides with no right angles. The two sets of parallel sides are of unequal lengths.] (b) The rhombus? [An equilateral parallelogram with no right angles.] (c) The trapezoid? [A quadrilateral with only two sides parallel. A quadrilateral is a figure in the plane bounded by four straight lines terminated at four angles.] For (c) divide the trapezoids into two sets: those with the two non-parallel sides of: (i) equal length and (ii) unequal length. Note that (c(i)) is the base section of an isosceles triangle. See the figure below:

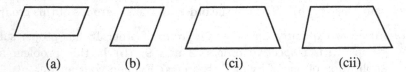

(a)   (b)   (ci)   (cii)

1.18 Describe a graphical means of determining the inverse of an element $X \in G$ from the group table of G.
Answer
With respect to the partial group table shown below we see that X and Y connect to E. As $XY = E$, it follows that $Y = X^{-1}$. [This

rule assists in the construction of classes of a given group (Section 2.2.)].

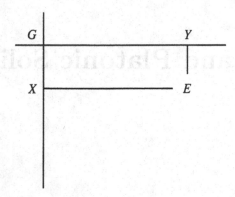

1.19 Show that any point group of prime order is isomorphic to the cyclic group of the same order.

1.20 An object in three-dimensional space has C_n rotational symmetry and one σ plane of reflection symmetry. Does the group of symmetries of this object include the S_n operation?
Answer
Only if the C_n axis is normal to the σ plane.

1.21 (a) What are the orders of the subgroups of a group, K, of order 15? (b) Show that these are the only subgroups of K. (c) Is this group unique?
Answers
(a) With Sylow's theorems and writing $15 = 3^1 \cdot 5^1$, we note that subgroups of K exist of order 3 and of order 5. (b) With Lagrange's theorem, we may conclude that these are the only subgroups of K. (c) As 3 and 5 are prime numbers, it follows that both subgroups are unique, whence K is unique. (That is, any other representation of K is isomorphic to it.)

2

Classes and Platonic Solids

2.1 Conjugate Elements

If A and X are elements of a group G, then $B = X^{-1}AX$ is also an element of G ($AX \equiv C \in G, X^{-1}C \in G$). In this event, one says that B *is conjugate to* A. It follows that every element of G is conjugate to some other element of G. Here are three important properties of conjugates:

(a) Every element of G is conjugate to itself,

$$X^{-1}XX = X. \tag{2.1}$$

(b) If A is conjugate to B, then B is conjugate to A,

$$A = X^{-1}BX,$$
$$XA = BX,$$
$$XAX^{-1} = B.$$

(c) Two elements conjugate to the same element are conjugate to each other. Consider that B and C are both conjugate with A,

$$A = X^{-1}BX, \tag{i}$$

$$A = Y^{-1}CY, \tag{j}$$

where all elements are contained in the group G. We then wish to show that

$$B = Z^{-1}CZ, \tag{k}$$

where $Z \in G$. From (i) and (j),

$$X^{-1}BX = Y^{-1}CY$$

so that

$$B = XY^{-1}CYX^{-1} = M^{-1}CM,$$

where

$$M \equiv YX^{-1} \in G$$

which establishes the theorem. In mathematics, the transformation $B = X^{-1}AX$ is called a *similarity transformation*.

2.2   Classes

A complete set of elements of a given group that are conjugate to each other is called a *class* of the given group. To demonstrate this concept, let us discover the classes of the sixth-order group C_{3v} whose group table is given by (1.10). As the operations σ_a, σ_b, σ_c are equivalent reflections, one expects that they are in the same class. Let us show this. With reference to (1.10) we find

$$
\begin{aligned}
E^{-1}\sigma_a E &= \sigma_a, & C_3^{-1}\sigma_a C_3 &= \sigma_c, \\
(C_3^2)^{-1}\sigma_a C_3^2 &= \sigma_b, & \sigma_a^{-1}\sigma_a \sigma_a &= \sigma_a, \\
\sigma_b^{-1}\sigma_a \sigma_b &= \sigma_c, & \sigma_c^{-1}\sigma_a \sigma_c &= \sigma_b.
\end{aligned}
\tag{2.2}
$$

The three σ elements form a closed set under similarity transformations with all elements of C_{3v}. It follows that one class of C_{3v} is $(\sigma_a, \sigma_b, \sigma_c)$. The remaining classes of C_{3v} are E, (C_3, C_3^2). Elements of a class of a given group are readily found with aid of the related group table.

Here are four properties of classes:

(a) The identity E is in a class by itself.

(b) An element of a group cannot be in more than one class.

(c) The number of elements in a class is an integral factor of the order of the group. An example of this property is given by the sixth-order group C_{3v}. Respective orders of the three classes 1, 2, 3 are all integral factors of 6. (The 'class-divisor theorem.')

(d) Every element of an Abelian group G is in a class by itself. ($ABA^{-1} = AA^{-1}B = B$ for any $B \in G$ and all remaining $A \in G$.)

(e) Elements of a class have common operator properties. Thus, as noted above, the collection of σ reflection operators relevant to a given group comprise a class of that group. The same is true, for example, for rotation operators C_n about a common axis.

2.3   Direct Product

Consider two groups, G_1 and G_2, with respective elements, $A_i \in G_1$ and $B_i \in G_2$, with the commuting property

$$A_i B_j - B_j A_i = 0 \tag{2.3}$$

The direct product of G_1 and G_2 is written, $G_1 \otimes G_2$, and is composed of the products $A_i B_j$. If the order of G_1 is h_1 and that of G_2 is h_2, then the order of $G_1 \otimes G_2$ is $h_1 h_2$. The elements of $G_1 \otimes G_2$ comprise a group. We demonstrate the closure property of $G_1 \otimes G_2$ and the property of the existence of inverses for respective elements of this group.

The closure property of $G_1 \otimes G_2$ is given by

$$(A_i B_j)(A_k B_n) = (A_i A_k)(B_j B_n) = A_s B_q \in G_1 \otimes G_2 \tag{2.4}$$

The inverse of an element $A_i B_j \in G_1 \otimes G_2$ is given by $B_j^{-1} A_i^{-1}$, namely,

$$(A_i B_j)^{-1}(A_i B_j) = B_j^{-1} A_i^{-1} A_i B_j = B_j^{-1} B_j = E \tag{2.5}$$

Note that the commutative property of elements of the direct product, (2.3), comes into play in (2.4).

Consider the direct product $C_2 \otimes C_3$. If elements of C_2 are (E, D) and those of C_3 are (E, A, B), then the direct product group is sixth order, with elements (E, A, B, D, DA, DB). This group is isomorphic to the Abelian group C_6.

Regarding classes of direct products, we note the following. If K_1 is a class of the group G_1 and K_2 is a class of G_2, then a class of the product group $G_1 \otimes G_2$ exists which is equal to $K_1 \otimes K_2$.

2.4 C_{nv} and D_n Groups

We wish to apply the C_{nv} and D_n groups to the symmetry of regular n-gon prisms. The general procedure will be illustrated for the case $n = 4$. A general compilation of properties of related point-group operations is presented in Table 1.1.

The C_{4v} and D_4 groups may be described with respect to a square prism. Elements of the C_{4v} group are composed of the eight operations

$$C_{4v} = (E, C_2, C_4, C_4^3, \sigma_{v1}, \sigma_{v2}, \sigma_{d1}, \sigma_{d2}) \tag{2.6}$$

In this and in following expressions, one makes the identification, $C_2 = C_4^2$. Consider a square cross section of the prism centered at the origin of (x, y) Cartesian axes in the plane, with edges parallel to these respective axes. Then the C_4 axis is the z axis (the principal axis of the prism). The reflection operations $(\sigma_{v1}, \sigma_{v2})$ are with respect to planes normal to the x, y

planes and through respective axes (Fig. 2.1) whereas the reflection operations $(\sigma_{d1}, \sigma_{d2})$ are with respect to planes through respective diagonals of the square (Fig. 1.5).

Elements of the C_{4v} group divide into the five classes

$$E; C_2; (C_4, C_4^3); (\sigma_{v1}, \sigma_{v2}); (\sigma_{d1}, \sigma_{d2}) \tag{2.7}$$

It proves convenient to rewrite this partitioning in the reduced form

$$E; C_2; 2C_4; 2\sigma_v; 2\sigma_d \tag{2.8}$$

which indicates that there are two C_4 operations in one class, etc.

The D_4 group is a pure rotation group whose elements are generated by the C_4 axis and the four C_2 axes perpendicular to C_4. With reference to the description of the square cross section, two of the four C_2 rotations are about the x and y axes, respectively, and two are about the diagonals of the square. Group elements are written

$$D_4 = [E, C_2, C_4, C_4^3, C_2(x), C_2(y), C_2(xy), C_2(x\bar{y})] \tag{2.9a}$$

where $\bar{y}$ denotes $-y$. The latter two rotations in (2.9a) are with respect to diagonals. In reduced notation, the elements (2.9) partition into the five classes

$$D_4 = [E, 2C_4, C_2, 2C_2', 2C_2''] \tag{2.9b}$$

where again it is noted that $C_2 = C_4^2$ and single and double primes relate to rotations about Cartesian and diagonal axes, respectively.

We recall the following properties of a regular polygon. For n even, a regular polygon has two sets of bisectors composed each of equivalent bisectors, whereas for n odd a regular polygon has only one set of equivalent bisectors. Both the D_n and C_{nv} point groups likewise partition according to whether n is eveness or odd.

D_n Groups

Consider a horizontal cross section of a regular polygon prism. In addition to C_n rotations about the principal axis, the D_n groups describe rotations about bisecting axes of this polygon. The D_{nh} groups include reflections in horizontal planes of the prism. Elements of the D_2 groups are given by

$$D_2 = [E, C_2(z), C_2'(x), C_2''(y)]$$

Here is a list of group elements of the D_n groups for $n = 3, \ldots, 6$, in reduced notation

$$D_3 = (E, 2C_3, 3C_2)$$
$$D_4 = (E, 2C_4, C_2, 2C_2', 2C_2'')$$
$$D_5 = (E, 2C_5, 2C_5^2, 5C_2)$$
$$D_6 = [E, 2C_6, 2C_3(= 2C_6^2), C_2(= C_6^3), 3C_2', 3C_2'']$$

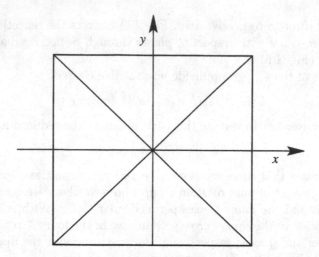

Figure 2.1. Symmetry configuration for the C_{4v} and D_4 groups. The C_{4v} group includes σ reflections through four planes that includes the principal axis and are at $\pi/4$ with one another. The D_4 group includes C_2 rotations about the four axes through the origin. Both groups include C_4 rotations about the z axis. Let the square figure be considered the face of a cube. In this event, symmetries generalize to three planes of reflection symmetry (σ_h) of the cube and the diagonal lines to six planes of reflection symmetry (σ_d) of the cube. See (2.13a).

Partitioning of the C_{nv} Groups

The application of the C_{nv} groups, to symmetries of the regular n-gon prisms, follows that for the $n = 4$ case, described above. These properties as well as properties of other related point groups are summarized in Table 2.1.

Table 2.1

Properties of Point Groups

I. The C_n group

This group describes symmetries of the regular polygons and includes n rotations through $2\pi/n$ about the central axis normal to the polygon. This group is Abelian. There are n elements in this group. The group $C_s = (E, \sigma_h)$.

II. The C_{nv} group

There are n vertical planes of a regular polygon prism that include the principal axis of the prism, bisect the prism and are at angles π/n with one another. The C_{nv} group includes the n elements of the C_n group. In addition, for n odd, there are n reflections, σ_v, through the n vertical planes.

For n even, there are an additional $n/2$ reflections, σ_d, and $n/2$ reflections, σ_v. There are $2n$ elements in this group.

III. The $\mathbf{C_{nh}}$ group

This group contains elements of the C_n group and the n elements obtained by multiplying elements of the C_n group by σ_h. This group is Abelian. If n is even, C_{nh} contains the inversion operator, i. There are $2n$ elements in this group.

IV. The $\mathbf{D_n}$ group

Consider a plane normal to the principal axis of a regular polygon prism that forms a regular polygon with n bisecting lines. For n odd, D_n includes n rotations through π about these bisecting lines. For n even, D_n includes $(n/2)$ rotations through π about bisecting lines through vertices of the polygon and $(n/2)$ rotations through π about bisecting lines through parallel sides of the polygon. In addition D_n contains the n elements of the C_n group. There are $2n$ elements in this group.

V. The $\mathbf{D_{nh}}$ group

This group contains the $2n$ elements of the D_n group and the $2n$ elements obtained by multiplying each element of the D_n group by σ_h. If n is even, D_{nd} includes the i operation. There are $4n$ elements in this group.

VI. The $\mathbf{D_{nd}}$ group

This group contains the $2n$ elements of the D_n group and the $2n$ elements obtained by multiplying each element of the D_n group by σ_d. If n is odd, D_{nd} includes the i operation. There are $4n$ elements in this group.

VII. The $\mathbf{S_n}$ group

The group S_{2n} contains $2n$ improper rotations. For n odd, $S_n = C_{nh}$. The group $S_2 = (E, S_2) = (E, i) = C_i$.

Here is list of the preceding properties.

Subgroup Relations

$$C_n \subset C_{nh}; \quad C_n \subset C_{nv}; \quad C_n \subset D_n$$

$$C_{nv} = C_n \cup (\sigma_v, \sigma_d); \quad D_{nd} = D_n \cup (\sigma_d D_n); \quad D_{nh} = D_n \cup (\sigma_h D_n)$$

$$T_d = T \cup (\sigma_d T); \quad T_h = T \cup (iT); \quad O_h = O \cup (iO).$$

We recall that the 'union' of two sets, $A \cup B$, represents the set of elements in A and B. The symbol $A \subset B$ stipulates that the set A is included in the

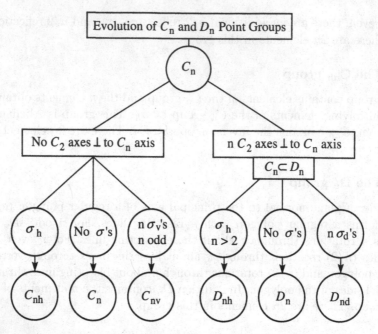

Figure 2.2. Evolution of C_n and D_n point groups.

set B. If A and B are groups, then $A \subset B$ indicates that A is subgroup of B. Reflection elements, (σ_v, σ_h), of the C_{nv} group, as noted in Table 2.1, depend on the eveness or oddness of n.

Commutator Relations

The following relations apply to the commutator (1.2) of two operators:

$$[C_n, C_{n'}] = 0 \tag{2.10a}$$

$$[\sigma_a, \sigma_b] = 0, \quad (\sigma_a \perp \sigma_b) \tag{2.10b}$$

$$[i, C_n] = 0 \tag{2.10c}$$

$$[i, \sigma] = 0 \tag{2.10d}$$

$$[\sigma, C_n] = 0 \ (\sigma \text{ reflects in a plane } \perp \text{ to } C_n \text{ axis}) \tag{2.10e}$$

In general, three operators A, B, C satisfy the relation (Jacobi's equation)

$$[A, [B, C]] + [B, [C, A]] + [C, [A, B]] = 0 \tag{2.10f}$$

The i Operation

We examine the inversion operator, i, for the C_{nh} and D_{nh} groups and show that for n even these groups include the i operation, and for n odd they do not. To show this we note that the i operation is composed of the σ_h and C_2 operations where C_2 is normal to a σ_h plane. Consider a regular n-gon prism. As σ_h reflects the prism through an h plane it leaves the prisim invariant. Now apply the C_2 operation. If n is odd (e.g., the equilateral triangle prism), the prism does not map onto itself, whereas if n is even (e.g., the square prism) it does. It follows that $i \in C_{nh}, D_{nh}$, only for n even.

Evolution of the C_n and D_n point groups is shown in Fig. 2.2.

2.5 Platonic Solids. T, O and I Groups

We have found that the regular n-gon prisms play a role in interpreting the point groups, $C_{nv}, C_{nh}, D_n, D_{nh}, D_{nd}$. The convex regular polyhedra in three dimensions play a role in the interpretation of higher order groups. A regular polyhedron has the following defining properties:

(a) Each face is a regular polygon, which are all equivalent.

(b) Angles of the vertices are all equal.

There are five such regular polyhedra: the tetrahedron, the cube, the octahedron, the dodecahedron and the icosahedron (Fig. 2.3). Let us establish that there are only five such regular polyhedra. We first examine the case that the polyhedron is composed of equilateral triangles. The following possibilities occur: 3, 4, or 5 triangles meet at a common pyramidal (non-planar) vertex. If 6 equilateral triangles meet at the vertex, then the angle about the vertex is $6 \times (\pi/3) = 2\pi$ and the vertex is planar. The three remaining possibilities correspond, respectively, to the tetrahedron, octahedron and the icosahedron.

For polyhedra composed of squares, only the case of three squares meeting at a vertex holds, which corresponds to the cube. (Four squares meeting at a common vertex lie in a plane.) Next consider the pentagon whose internal angles equal $3\pi/5$. For three such pentagons meeting at a common vertex, the angle about the vertex is $9\pi/5 < 2\pi$, which is pyramidal and corresponds to the dodecahedron.

The internal angles of a hexagon are $2\pi/3$. For three such hexagons meeting at a common vertex, the angle about the vertex is $3(2\pi/3)$ which corresponds to a plane surface. We conclude that the surfaces of regular polyhedra are comprised, respectively, only of equilateral triangles, squares and pentagons and that there are only five regular polyhedra.

Another property of regular polyhedra important to group theoretical arguments is that of the *dual*. Consider a regular convex polyhedron, A, with n vertices. Let n radii from the geometric center of A pass, respectively, through each of its n vertices. Let $\mathbf{r}$ be one such radius. Let a plane pass normally to $\mathbf{r}$ at the value $1/r$. The envelope of n such planes is the *dual*

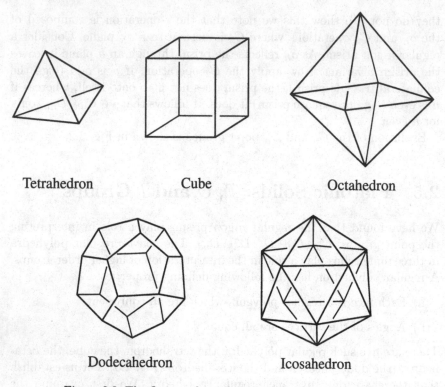

Tetrahedron   Cube   Octahedron

Dodecahedron   Icosahedron

Figure 2.3. The five regular polyhedra. The Platonic solids.

of A. Let B be the dual of A. It follows that vertices of A map onto faces of B and faces of A map onto vertices of B. We note that:

(a) The tetrahedron is self-dual.

(b) The cube is dual to the octahedron.

(c) The dodecahedron is dual to the icosahedron.

(d) The same group of symmetries apply to a polyhedron and its dual.

(e) A polyhedron and its dual have the same number of bisecting surfaces of symmetry.

(f) A polyhedron and its dual have the same number of edges.

Properties of the regular polyhedra are listed in Table 2.2.

Table 2.2

Properties of Regular Polyhedra

Name	Faces	Vertices	Edges	Bisecting Surfaces	Group and Order
Tetrahedron	4 equilateral triangles	4	6	6	T_d (24)
Cube	6 squares	8	12	9	O_h (48) (Dual)
Octahedron	8 equilateral triangles	6	12	9	O_h (48)
Dodecahedron	12 pentagons	20	30	15	I_h (48) (Dual)
Icosahedron	20 equilateral triangles	12	30	15	I_h (48)

Point Groups of the Regular Polyhedra

(a) The T Groups

Symmetries of the tetrahedron are given by the T_d group of order 24 with elements (in reduced notation)

$$T_d = (E, 8C_3, 3C_2, 6S_4, 6\sigma_d) \tag{2.11a}$$

The tetrahedron does not have a center of inversion symmetry so that $i \notin T_d$. The $8C_3$ term refers to $2\pi/3$ rotations about eight axes of rotational C_3 symmetry of the tetrahedron. This group has a purely rotational subgroup, T, of order 12, with elements

$$T = (E, 4C_3, 4C_3^2, 3C_2) \tag{2.11b}$$

The T_d group is then given by the elements of T and products of T with σ_d,

$$T_d = (T, \sigma_d T) \tag{2.11c}$$

The planes of σ_d contain one C_2 axis and bisect the other two. With $C_s = (E, \sigma_d)$, T_d may be written as the direct product

$$T_d = T \otimes C_s \tag{2.11d}$$

A closely allied group, T_h, is obtained by adding to the elements of T, the products iT,

$$T_h = (T, iT). \tag{2.12a}$$

Recalling the group C_i (1.7), permits T_h to be written as the direct product

$$T_h = T \otimes C_i. \tag{2.12b}$$

(b) The O Groups

Symmetries of the cube and octahedron are given by the O_h group of order 48 with elements

$$O_h = [E, 8C_3, 6C_4, 6C_2, 3C_2 \, (= C_4^2), i, 6S_4, 8S_6, 3\sigma_h, 6\sigma_d]. \tag{2.13a}$$

These polyhedra have respective centers of inversion symmetry, so that $i \in O_h$. This group has a purely rotational subgroup, O, of order 24 with elements

$$O = [E, 8C_3, 6C_4, 6C_2, 3C_2 \, (= C_4^2)]. \tag{2.13b}$$

The group O_h may be written as the direct product

$$O_h = O \otimes C_i. \tag{2.13c}$$

(c) The I Groups

Symmetries of the icosahedron and the dodecahedron are given by the I_h group of order 120 with elements

$$I_h = [E, 12C_5, 12C_5^2, 20C_3, 15C_2, i, 12S_{10}, 12S_{10}^3, 12S_6, 15\sigma]. \tag{2.14a}$$

These polyhedra likewise have respective centers of inversion symmetry, so that $i \in I_h$. This group has a purely rotational subgroup, I, of order 60 with elements

$$I = (E, 12C_5, 12C_5^2, 20C_3, 15C_2) \tag{2.14b}$$

which is related to the group I_h through the direct product

$$I_h = I \otimes C_i. \tag{2.15a}$$

The groups C_n and S_n are both cyclic. For n odd, $S_n = C_{nh}$. For n even, $C_{n/2} \subset S_n$ and $S_n^n = E$. The related cyclic group is easily constructed from powers of S_n. For example,

$$\begin{aligned} S_6 &= (S_6, S_6^2, S_6^3, S_6^4, S_6^5, E) = (S_6, C_6^2, S_6^3, C_6^4, S_6^5, E) \\ &= (S_6, C_3, i(= S_2), C_3^2, S_6^5, E) \supset (E, C_3, C_3^2) = C_3. \end{aligned} \tag{2.15b}$$

h-Group Direct Products

The former relations involving h-groups and related groups are included in the following list of direct products.

$$C_{nh}, C_n, D_n \text{ and } S_n \text{ direct-product groups}$$

$$C_{3h} = C_3 \otimes C_S, \qquad D_{2h} = D_2 \otimes C_i,$$

$$C_{4h} = C_4 \otimes C_i, \qquad D_{4h} = D_4 \otimes C_i,$$

$$C_{6h} = C_6 \otimes C_i, \qquad D_{6h} = D_6 \otimes C_i,$$

$$S_6 = C_3 \otimes C_i, \qquad D_{3d} = D_3 \otimes C_i,$$

$$\text{The Platonic Groups}$$

$$T_h = T \otimes C_i, \qquad O_h = O \otimes C_i,$$

$$T_d = T \otimes C_s, \qquad I_h = I \otimes C_i.$$

The group $S_2 = (E, S_2) = (E, i) = C_i$ and $C_s(E, \sigma)$, where the σ plane relates to the group at hand. For example, in D_{nh}, $C_s = (E, \sigma_h)$.

We recall that the direct product is defined with respect to commuting groups. For consistency of the preceding relations we must show that i and σ, respectively, commute with elements of related groups. Consider the product (iF) where F denotes an element of a point group. Let a point in the domain of F be labeled (x, y, z). Consider the term

$$iF(x, y, z) = i(x', y', z') = (-x', -y', -z') = F(-x, -y, -z) = Fi(x, y, z) \tag{2.16a}$$

So that $C_i F = F C_i$. With regard to the product σF, with no loss in generality, we assume that the σ operator reflects through the $x = 0$ plane. There results

$$\sigma F(x, y, z) = \sigma(x', y', z') = (-x', y', z') = F(-x, y, z) = F\sigma(x, y, z) \tag{2.16b}$$

so that $\sigma F = F\sigma$. Group and geometrical properties of the regular polyhedra are listed in Table 2.2.

The Cubic Groups

The groups T, T_d, T_h $(= T \otimes S_2)$, O and O_h $(= O \otimes S_2)$ are called *cubic groups*. The T groups pertain to systems which contain three equivalent principal axes and the O groups pertain to systems which contain three principal axes and three symmetry axes, such as the cube. Furthermore, T_d is a subgroup of O_h. Note the subgroup relations

$$T \subset T_d \subset O_h; \quad O \subset O_h. \tag{2.16c}$$

The groups T and O are purely rotational.

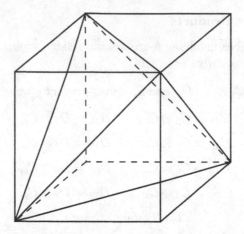

Figure 2.4. Tetrahedron inscribed in a cube. Each edge of the tetrahedron lies on a diagonal of a cube face.

The Tetrahedron and the Cube

As shown in Fig. 2.4, a tetrahedron may be inscribed in a cube such that each edge of the tetrahedron lies on a diagonal of a cube face. The T_d group contains C_2 rotations about axes perpendicular to each of the three pairs of cube faces, which gives the class $3C_2$ as well as 6 reflections though planes which pass normally through the 6 faces of a cube along which the 6 edges of the inscribed tetrahedron lie (Fig. 2.4) which give the $6\sigma_a$ class. It also contains four C_3 rotations about the cube diagonals (through opposite vertices of the cube) and four corresponding C_3^2 rotations giving the class $8C_3$ thus there are 24 elements of the T_d group, (2.11a).

Non-Regular Dual Polyhedra

We note the following theorem. If (X, Y) represent a set of dual polyhedra, then $[X(\text{stellated}), Y(\text{truncated})]$ are likewise a set of dual polyhedra, and described by the same group as (X, Y). We recall that in the stellation of a regular polygon respective centers of the polygon edges are displaced normally from the edge. Thus the stellation of a pentagon is a pentagram (Fig. 2.5). The truncation of a regular polygon is obtained by a relatively small inward displacement of respective centers of polygon edges. Corresponding to the Platonic regular polyhedra there are three pairs of complimentary dual polyhedra. These are listed in Table 2.3 together with group properties and names of polyhedra. In this process, truncation of vertices and stellation of faces are uniform. Note that a stellated regular convex polyhedron is non-convex. As the tetrahedron is self-dual these two operations on a tetrahedron give two dual polyhedra.

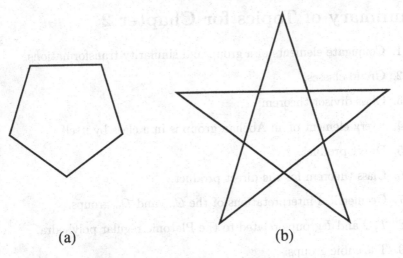

(a)   (b)

Figure 2.5. The pentagon and, (b) its stellation, the pentagram.

Table 2.3

Non-regular Dual Polyhedra

Name	(vertices, faces, edges)	Group
Truncated tetrahedron	(12, 8, 18)	T_d
Stellated tetrahedron	(8, 12, 18)	
Truncated octahedron	(24, 14, 36)	O_h
Stellated cube	(14, 24, 36)	
Cuboctahedron	(12, 14, 24)	I_h
Rhombohedral dodecahedron	(14, 12, 24)	

Cyclic and Abelian Groups

In concluding this chapter, the following is noted: Let $G(A)$ denote the fact that the group G is Abelian and let $G(C)$ denote the fact that G is cyclic. Then

$$G(C) \Rightarrow G(A)$$

$$G(A) \not\Rightarrow G(C)$$

That is, any cyclic group is Abelian. But an Abelian group is not necessarily cyclic. For example, C_n is cyclic and Abelian, but C_{2v} (1.9) which is Abelian is not cyclic. In either case, every element of an Abelian or cyclic group is in a class by itself.

Summary of Topics for Chapter 2

1. Conjugate elements of a group and similarity transformations.

2. Group classes.

3. Class divisor theorem.

4. Every element of an Abelian group is in a class by itself.

5. Direct product.

6. Class theorem for the direct product.

7. Geometrical interpretations of the C_{nv} and D_n groups.

8. T, O and I groups related to the Platonic regular polyhedra.

9. The cubic groups.

10. Non-regular dual polyhedra.

Problems

2.1 Show that the classes of C_{3v}, in addition to the σ class given by (2.2), are E and (C_3, C_3^2).

2.2 Show that every element of an Abelian group is in a class by itself.

2.3 Show that an element of a group cannot be in two classes of that group.

2.4 What are the classes of the product group, $C_2 \otimes C_3$?

2.5 In what manner are the C_{4v}, D_4 groups related? (Words only.)

2.6 Do the groups C_{3v} and C_{4v} have a direct product? Justify your answer. If your answer is yes, cite a class of this direct product.

2.7 Consider a finite group of order 6. Show that the respective subgroups of order 2 and 3 are conjugate to each other. Cite the theorem which establishes the validity of this property.

2.8 Consider a plane which intersects a square prism normal to the principal axis of the prism resulting in a square. (a) Show that the C_4 operation maintains the sense of rotational order of the square whereas σ_d or σ_v reverse this sense of rotation. (b) What effect does the inversion operator, i, have on the sense of rotation? *Hint:* Consecutively label corners of the square (in counterclockwise direction) a, b, c, d.

2.9 Show that all elements of a class of a given group have the same eigenvalues.

2.10 Show that eigenvectors and eigenvalues of a given operator are preserved under a similarity transformation.

2.11 (a) What are the group of symmetries of the rectangular slab of dimensions $a < b < c$, where a is the thickness of the slab? (b) List the elements of this group.
Answers (partial)
(a) The D_h group. (Let the rectangular slab be centered at the origin of a Cartesian frame with its bisecting surface of the a dimension on the $z = 0$ plane.)

2.12 The H_2O molecule in liquid water is a dipole with the $2H^+$ ions corresponding to the positive end and the O^- ion to the negative end. When NaCl dissolves in water, the molecule dissociates into Na^+ and Cl^- ions. The Na^+ ion is surrounded by six H_2O dipoles. If the resulting configuration of minimum energy of the six H_2O dipoles is a regular array, what group of symmetries describes this cluster?

2.13 The C_{nv} group contains no C_2 elements for n odd. The D_{nv} group contains C_2 elements, both for even and odd n. Explain this difference.

Answer

Rotations in C_{nv} are parallel to the principal axis of a regular n-gon prism. It follows that the C_{nv} group (n odd) contains no C_2 elements. The C_2 rotations in the D_{nv} group are about axes running through opposite vertices or faces (n even) of a regular n-gon (prism cross section) or through a vertex and an opposite face (n odd), for which C_2 always exists.

2.14 (a) Write down the group table of the product group $C_2 \otimes C_3$ where $C_2 = (E, C_2)$, $C_3 = (E, C_2, C_3)$. (b) Identify a group which is isomorphic to this product group.

2.15 In the direct-product group

$$G = A_1 \otimes A_2$$

show that

$$A_1 \subset G, A_2 \subset G.$$

2.16 The surface of a soccer ball is composed of uniformly distributed pentagons each surrounded by five hexagons. (a) What is the point group of a soccer ball with a minimum number of flat sides? (b) How many sides does this soccer ball have?

Answer (partial)

From the given description, a soccer ball is a truncated icosahedron. With Table 2.3, we conclude that the point group of a soccer ball is the I_h group.

2.17 If $k < n$ are integers and k divides n, show that the rotation operators C_n and S_n obey the relations

$$C_n^k = C_{n/k}; \qquad S_n^k = S_{n/k}$$

2.18 Euler's formula for a polyhedron with V vertices, E edges and F faces is given by

$$V - E + F = 2$$

Employ this relation to show that there are five regular polyhedra.

Answer

Each face of the regular polyhedron is a regular n-gon with n edges. Each edge joins two faces. It follows that

$$nF = 2E$$

Each vertex includes r edges and each edge joins two vertices so that

$$rV = 2E$$

Substituting these relations into Euler's formula gives

$$\frac{1}{r} + \frac{1}{n} = \frac{1}{2} + \frac{2}{nF}$$

The regular polyhedra correspond to the positive integer (n, r) solutions of this relation. Since each regular n-gon has $E > 3$ edges and each vertex has $r \geq 3$ edges we examine $n = 3; r = 3, 4, 5$ which give

$$F = 4, 8, 20$$

For $n = 4, 5; r = 3$ we obtain

$$F = 6, 12$$

These five face numbers correspond to the five regular polyhedra (see Table 2.2).

3

Matrices, Irreps and the Great Orthogonality Theorem

3.1   Matrix Representations of Operators

Basic Matrix Properties

Character of a Matrix

The character of a matrix (the "trace" in physics), which carries the symbol χ, is equal to the sum of the diagonals of the matrix. For the matrix A_{ij} we write

$$\chi = \sum_j A_{jj} \tag{3.1}$$

Here are two important properties of the character of a matrix:

Theorem: Let AB denote the product of two square matrices, A and B, of common dimension. Then

$$\chi_{AB} = \chi_{BA} \tag{3.2}$$

Here is the proof of this property

$$\chi_{AB} = \sum_n \sum_j A_{nj} B_{jn} = \sum_n \sum_j B_{jn} A_{nj} = \chi_{BA} \tag{3.3}$$

Theorem: If R and P are conjugate elements of a group (i.e., $R = Q^{-1}PQ$), then $\chi_R = \chi_P$. To prove this relation we write $\chi_R \equiv \chi(R)$. It follows that

$$\chi(R) \quad = \quad \chi(Q^{-1}PQ) = \chi[(Q^{-1}P)Q]$$

$$= \chi[Q(Q^{-1}P)] = \chi[QQ^{-1}P] = \chi(P). \tag{3.4}$$

Block-Partitioned Matrices

Consider two matrices of equal dimensions composed of block matrices along the diagonal. Parallel block matrices are likewise of equal dimension. The product of these two matrices is also block diagonal with dimensions of block diagonal matrices maintained. An example of this property is given below.

$$\begin{pmatrix} [2 \times 2(a)] & & \\ & [1 \times 1(b)] & \\ & & [3 \times 3(c)] \end{pmatrix} \begin{pmatrix} [2 \times 2(a')] & & \\ & [1 \times 1(b')] & \\ & & [3 \times 3(c')] \end{pmatrix}$$

$$= \begin{pmatrix} [2 \times 2(a,a')] & & \\ & [1 \times 1(b,b')] & \\ & & [3 \times 3(c,c')] \end{pmatrix} \tag{3.5}$$

where, for example, $2 \times 2(a,a')$ denotes a 2×2 matrix that is a function of (a,a').

Basic Matrix Definitions

(a) The *inverse*, M^{-1}, of a matrix M is given by

$$M^{-1} = \frac{1}{|M|} \widetilde{M}_c$$

where $|M|$ is written for the determinant of M and the matrix M_c consists of the cofactors of M. The *transpose* $\widetilde{M}$ of the matrix M is obtained by reflecting elements of M through the diagonal of M. That is,

$$\widetilde{M}_{ij} = M_{ji}$$

(The transpose of M is also written M^T.)

(b) A matrix M is *non-singular* iff M^{-1} exists. [With (a) it follows that a matrix M is *singular* iff $|M| = 0$.] (a) A real matrix is orthogonal iff

$$M^{-1} = \widetilde{M}, \quad \widetilde{M}M = I$$

where I is the identity matrix.

(d) The *Hermitian adjoint* of a matrix M is written $M^\dagger$ and is given by

$$M^\dagger = (\widetilde{M})^*$$

That is, $M^\dagger$ is the complex conjugate of $\widetilde{M}$. If M is real, then $M^\dagger = \widetilde{M}$.

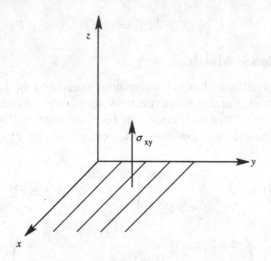

Figure 3.1. The σ_{xy} operation.

(d) A matrix M is *Hermitian* iff $M^\dagger = M$.

(e) A matrix is *unitary* iff

$$M^\dagger = M^{-1}$$

In this case

$$M^\dagger M = I$$

Application of these theorems and properties follows:

Matrices of Elementary Geometric Transformations in Three Dimensions.

Identity Matrix

The identity matrix is given by the unit diagonal matrix which, when operating on a vector (x, y, x), leaves it invariant,

$$\begin{pmatrix} 1 & 0 & 0 \\ 0 & 1 & 0 \\ 0 & 0 & 1 \end{pmatrix} \begin{pmatrix} x \\ y \\ z \end{pmatrix} = \begin{pmatrix} x \\ y \\ z \end{pmatrix} \tag{3.6}$$

Reflection Matrices

The σ_{xy} matrix reflects through the $z = 0$ plane as shown in Fig. 3.1.
The corresponding matrix equation is given by

$$\overset{\sigma_{xy}}{\begin{pmatrix} 1 & 0 & 0 \\ 0 & 1 & 0 \\ 0 & 0 & -1 \end{pmatrix}} \begin{pmatrix} x \\ y \\ z \end{pmatrix} = \begin{pmatrix} x \\ y \\ -z \end{pmatrix} \tag{3.7}$$

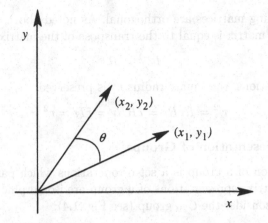

Figure 3.2. The rotational transformation.

Matrix equations of the remaining two reflection matrices are given by

$$\sigma_{xz}$$

$$\begin{pmatrix} 1 & 0 & 0 \\ 0 & -1 & 0 \\ 0 & 0 & 1 \end{pmatrix} \begin{pmatrix} x \\ y \\ z \end{pmatrix} = \begin{pmatrix} x \\ -y \\ z \end{pmatrix} \tag{3.8}$$

$$\sigma_{yz}$$

$$\begin{pmatrix} -1 & 0 & 0 \\ 0 & 1 & 0 \\ 0 & 0 & 1 \end{pmatrix} \begin{pmatrix} x \\ y \\ x \end{pmatrix} = \begin{pmatrix} -x \\ y \\ z \end{pmatrix} \tag{3.9}$$

$$i \text{ (inversion)}$$

$$\begin{pmatrix} -1 & 0 & 0 \\ 0 & -1 & 0 \\ 0 & 0 & -1 \end{pmatrix} \begin{pmatrix} x \\ y \\ z \end{pmatrix} = \begin{pmatrix} -x \\ -y \\ -z \end{pmatrix} \tag{3.10}$$

Proper Rotations

The coordinate z is unchanged by proper rotation about the z axis. Consider the vector (x_1, y_1) in the xy plane, rotated about the origin (counter-clockwise) through the angle θ, to the vector (x_2, y_2), as shown in Fig. 3.2.

The corresponding matrix relation is given by

$$\begin{pmatrix} \cos\theta & -\sin\theta & 0 \\ \sin\theta & \cos\theta & 0 \\ 0 & 0 & 1 \end{pmatrix} \begin{pmatrix} x_1 \\ y_1 \\ z_1 \end{pmatrix} = \begin{pmatrix} x_2 \\ y_2 \\ z_1 \end{pmatrix} \tag{3.11}$$

All the preceding matrices are orthogonal. As noted above, the inverse of an orthogonal matrix is equal to the transpose of the matrix,

$$R^{-1} = \widetilde{R} \tag{3.12}$$

For such operations, the square radius r^2 is preserved,

$$r'^2 = \widetilde{Rr}\,Rr = \tilde{r}\widetilde{R}Rr = \tilde{r}Ir = r^2 \tag{3.13}$$

Matrix Representation of Groups

A representation of a group is a set of operations which parallel those of the group. Matrix representations of a group are isomorphic to the group. For example, consider the C_{2v} group (see Fig. 1.4):

$$C_{2v} = (E, C_2, \sigma_v, \sigma_v')$$

Corresponding matrices are given by the forms

$$E: \begin{pmatrix} 1 & 0 & 0 \\ 0 & 1 & 0 \\ 0 & 0 & 1 \end{pmatrix}, \quad C_2: \begin{pmatrix} -1 & 0 & 0 \\ 0 & -1 & 0 \\ 0 & 0 & 1 \end{pmatrix},$$

$$\sigma_v: \begin{pmatrix} 1 & 0 & 0 \\ 0 & -1 & 0 \\ 0 & 0 & 1 \end{pmatrix}, \quad \sigma_v': \begin{pmatrix} -1 & 0 & 0 \\ 0 & 1 & 0 \\ 0 & 0 & 1 \end{pmatrix} \tag{3.14}$$

These matrices have the group table given by (1.9) where superscripts on σ are now written as primes. For example, consider the product

$$\sigma_v C_2 = \sigma_v'$$

$$\begin{pmatrix} 1 & 0 & 0 \\ 0 & -1 & 0 \\ 0 & 0 & 1 \end{pmatrix} \begin{pmatrix} -1 & 0 & 0 \\ 0 & -1 & 0 \\ 0 & 0 & 1 \end{pmatrix} = \begin{pmatrix} -1 & 0 & 0 \\ 0 & 1 & 0 \\ 0 & 0 & 1 \end{pmatrix}$$

The four matrices (3.14) comprise an isomorphic representation of C_{2v}. Here are four examples of homomorphic representations of C_{2v}:

	E	C_2	σ_v	σ_v'
Γ_1	1	1	1	1
Γ_2	1	-1	1	-1
Γ_3	1	-1	-1	1
Γ_4	1	1	-1	-1

$$\tag{3.15}$$

The four Γ representations are each homomorphic to the C_{2v} group. Each Γ representation has a multiplication table with parallel mappings to that of the C_{2v} group. In the construction of group representations, the rearrangement theorem is often lost. This occurs when elements of such groups are not unique as is the case for homomorphic maps for which representations are composed only of 1's such as is the case in the preceding example.

Similarity Transformations

An important group representation stems from similarity (or 'conjugation') transformations of group elements. Consider an operator, $Q \notin G$, which exists in the same binary space of operations as the elements of a group G. Elements of the similarity group, G', are given by the transformation

$$E' = Q^{-1}EQ$$
$$A' = Q^{-1}AQ \qquad (3.16)$$
$$\vdots$$

Thus, if $AB = D$ in G, then

$$A'B' = (Q^{-1}AQ)(Q^{-1}BQ) = Q^{-1}ABQ = Q^{-1}DQ = D'$$

and $A'B' = D'$ in G'. Elements of G are *conjugated* to elements of G' by the operator Q.

3.2   Irreducible Representations

Again let the group G have elements $A, B, C, \ldots$. Consider the case that when the matrix representation of A is conjugated to the matrix A' by a given matrix Q, a block diagonal matrix is generated, composed of square matrices as shown below,

$$A' = Q^{-1}AQ \rightarrow \begin{pmatrix} A'_1 & 0 & 0 & 0 & 0 \\ 0 & A'_2 & 0 & 0 & 0 \\ 0 & 0 & . & 0 & 0 \\ 0 & 0 & 0 & . & 0 \\ 0 & 0 & 0 & 0 & \end{pmatrix} \qquad (3.17)$$

The dimension of the A'_1 matrix is $\ell_1 \times \ell_1$ and the dimension of the A'_2 matrix is $\ell_2 \times \ell_2$ and so forth. Similarly for the B element we write

$$B' = Q^{-1}BQ \rightarrow \begin{pmatrix} B'_1 & 0 & 0 & 0 & 0 \\ 0 & B'_2 & 0 & 0 & 0 \\ 0 & 0 & . & 0 & 0 \\ 0 & 0 & 0 & . & 0 \\ 0 & 0 & 0 & 0 & \end{pmatrix} \qquad (3.18)$$

The dimension of the B'_1 matrix is likewise $\ell_1 \times \ell_1$ and the dimension of the B'_2 matrix is $\ell_2 \times \ell_2$ and so forth.

Finally we write

$$E' = Q^{-1}EQ \rightarrow \begin{pmatrix} E'_1 & 0 & 0 & 0 & 0 \\ 0 & E'_2 & 0 & 0 & 0 \\ 0 & 0 & . & 0 & 0 \\ 0 & 0 & 0 & . & 0 \\ 0 & 0 & 0 & 0 & \end{pmatrix} \qquad (3.19)$$

where E'_1 is an $\ell_1 \times \ell_1$ unit matrix and E'_2 is an $\ell_2 \times \ell_2$ unit matrix and so forth.

If $AB = D$ in G, then multiplication of the matrices (3.12) et seq., gives consistent group products

$$A'_1 B'_1 = D'_1, \quad A'_2 B'_2 = D'_2, \quad A'_3 B'_3 = D'_3, \text{ etc.} \qquad (3.20)$$

In this manner one obtains the following representations of G:

$$\text{Rep } \Gamma_1: A'_1, B'_1, \ldots, E'_1$$
$$\text{Rep } \Gamma_2: A'_2, B'_2, \ldots, E'_2$$
$$\text{Rep } \Gamma_3: A'_3, B'_3, \ldots, E'_3 \qquad (3.21)$$

Suppose it is the case that the dimensions $\ell_1, \ell_2, \ldots$ of the G_i submatrices cannot be further reduced by similarity transformations. Then the representations $\Gamma_1, \Gamma_2, \ldots$ are said to be *irreducible representations* ('irreps') of the group G. Note that all one-dimensional representations are irreducible [see (3.15)].

3.3   Great Orthogonality Theorem (GOT)

Notation

1. We recall that h represents the order of a group.

2. The dimension of the ith component of each irrep is written ℓ_i.

3. A group operation is represented by the symbol R.

4. The mth row and nth column of the matrix corresponding to the operation R in the ith irrep is denoted by $\Gamma_i(R)_{mn}$.

With these preliminaries the Great Orthogonality Theorem (GOT) is written

$$\sum_R [\Gamma_i(R)_{mn}][\Gamma_j(R)_{m'n'}]^* = \frac{h}{\sqrt{\ell_i \ell_j}} \delta_{ij} \delta_{mm'} \delta_{nn'} \qquad (3.22a)$$

The sum is over all operations R at fixed $n, n'; m, m'; i, j$. Note in particular that the only way for this sum to survive is for $i = j, m = m', n = n'$.

The general structure of a diagonalized group $G(R)$ is given by

$$QG(R)Q^{-1} = \begin{pmatrix} \boxed{} & & \\ & \boxed{\Gamma_i(R)} & \\ & & \boxed{} \end{pmatrix} \qquad (3.22b)$$

Examples

Examples of the workings of the GOT are well demonstrated by the following three irreps of C_{3v}, for which $h = 6$:

C_{3v}	E	C_3	C_3^2	σ_v	σ_v'	σ_v''
Γ_1	1	1	1	1	1	1
Γ_2	1	1	1	-1	-1	-1
Γ_3	M_1	M_2	M_3	M_4	M_5	M_6

$$(3.23)$$

where

$$M_1 = \begin{pmatrix} 1 & 0 \\ 0 & 1 \end{pmatrix}, \qquad M_2 = \frac{1}{2}\begin{pmatrix} -1 & \sqrt{3} \\ \sqrt{3} & -1 \end{pmatrix},$$

$$M_3 = \frac{1}{2}\begin{pmatrix} -1 & \sqrt{3} \\ -\sqrt{3} & -1 \end{pmatrix}, \qquad M_4 = \begin{pmatrix} -1 & 0 \\ 0 & 1 \end{pmatrix}, \qquad (3.24)$$

$$M_5 = \frac{1}{2}\begin{pmatrix} 1 & \sqrt{3} \\ -\sqrt{3} & -1 \end{pmatrix}, \qquad M_6 = \frac{1}{2}\begin{pmatrix} 1 & \sqrt{3} \\ \sqrt{3} & -1 \end{pmatrix}.$$

These M_j elements comprise a matrix representation of the C_{3v} group and include the terms $\sin(2\pi/3)$ and $(\cos 2\pi/3)$, relevant to the C_3 rotations.

In the following, we denote group elements in the first column of the list (3.23) as R, the second column as R', etc. In the first example we set all indices in (3.22) to be equal. There results

$$[\Gamma_1(R)_1]^2 + [\Gamma_1(R')_1]^2 + \ldots = \frac{h}{\sqrt{1}} 1 \times 1 \times 1 = 6. \qquad (3.25)$$

As noted above, this is the only situation for which the sum in (3.17) survives. For our second example we choose the values $i = 1$, $j = 3$ and $m = n = 2$. There results

$$\Gamma_1(R)_{22}\Gamma_3(R)_{22} + \Gamma_1(R')_{22}\Gamma_3(R')_{22} + \ldots = 1 - \frac{1}{2} - \frac{1}{2} + 1 - \frac{1}{2} - \frac{1}{2} = 0 \quad (3.26)$$

which is a demonstration of the orthogonality property of the GOT.

The following six rules are related to the GOT and the theory of irreducible representations.

3.4   Six Important Rules

For purposes of discussion of the following rules, consider the table of characters of the C_{3v} group. With reference to (3.23) and writing classes in reduced form (Section 2.4), the following *character table* is obtained (discussed more fully in Section 3.5):

$$
\begin{array}{c|ccc}
C_{3v} & E & 2C_3 & 3\sigma_v \\
\hline
\Gamma_1 & 1 & 1 & 1 \\
\Gamma_2 & 1 & 1 & -1 \\
\Gamma_3 & 2 & -1 & 0
\end{array}
\tag{3.27}
$$

I. The sum of the squares of the dimensions of the irreps of a group is equal to the order of the group,

$$
\sum_i^{N(\text{irrep})} \ell_i^2 = h
\tag{3.28}
$$

The summation is over all irreps of a given group of order h. For the C_{3v} group of order 6, the preceding rule indicates that the irreps in (3.27) are the only irreps of this group. [Note that the sum (3.28) is over any column in a character table. In the $2C_3$ column, each irrep is composed of one-dimensional matrices.] We label '2' for this case, the 'class-multiplicity factor.'

II. As the character of the identity matrix is equal to the dimension of the matrix, with (3.22) we obtain

$$
\chi_i(E) = \ell_i, \qquad \sum_i [\chi_i(E)]^2 = h
\tag{3.29}
$$

III. The characters of irreps of a group act as a set of orthonormal vectors,

$$
\frac{1}{h} \sum_R \chi_i(R)\chi_j(R) = \delta_{ij}
\tag{3.30}
$$

The sum is over two parallel rows in a character table.

IV. The characters of all elements of the same class of a group are equal.

V. The number of irreps of a group equals the number of classes of the group. For example, we recall that the group C_{3v} has three classes. It follows that (as found above) this group has three irreps.

VI. The number of elements in a class is a divisor of h. Again with reference to the group C_{3v} we note the classes $E, 2C_3$ and $3\sigma_v$, whose class numbers 1, 2, 3, all divide 6, the order of C_{3v}.

Here are some applications of rule **III**. First consider the Γ_3 irrep for which, with $i = j = 3$, gives

$$
2^2 + 2(-1)^2 + 3(0) = 6
\tag{3.31}
$$

Consider next, application of the orthogonality component of this rule to the irreps Γ_1 and Γ_3 of the group (3.26). There results

$$\sum_R \chi_1(R)\chi_3(R) = (1) \times (2) + 2(1) \times (-1) + 3(1) \times (0) = 0 \qquad (3.32)$$

Another application of rule **III** is as follows. In a character table, such as that given by (3.22), the first row always depicts the one-dimensional irrep, labeled A_1. This row appears as

G	E	N_1x	N_2x'	N_3x''
A_1	1	1	1	1

where $x, x'x'', \ldots$, denote group elements in respective classes and $N_1, N_2, \ldots$, denote the number of elements in related classes. Using rule **III** with $i = j = 1$, we may write

$$1 + N_1 + N_2 + \ldots = h \qquad (3.33)$$

This equation may be read from any character table (Section 3.5).

Three Related Theorems

Three important theorems related to the GOT are as follows:

Theorem 1. For any matrix representation of a given group, a non-singular matrix exists which by similarity transforms, maps elements of that representation into a unitary matrix representation (i.e., a representation consisting only of unitary matrices).

Theorem 2 (Schur's Lemma). Let

$$\Gamma(R)M = M\Gamma(R)$$

for all R in the group representation Γ and let M be a matrix of dimension equal to that of Γ. Then: (a) If $\Gamma(R)$ is an irrep, M is a constant (i.e., a constant times an identity matrix). (b) If M is not a constant, then $\Gamma(R)$ is reducible.

Theorem 3. Let $\Gamma_i(R)$ and $\Gamma_j(R)$ be two irreps of a group G. The two irreps have respective dimensions p and q. Consider a rectangular matrix M of dimension $p \times q$ with the property $M\Gamma_i(R) = \Gamma_j(R)M$ for any $R \in G$. Then: (a) If $p \neq q, M = 0$ (all elements of M are zero). (b) If $p = q$, M is a square matrix and $|M| \neq 0$. In this event, the two representations are equivalent.

Example

Let us employ the preceding rules to rediscover the characters of the irreps of the C_{3v} group listed in (3.23). As the order of this group is $h = 6$ and the number of classes is 3, with rules **I** and **V** we write

$$\ell_1^2 + \ell_2^2 + \ell_3^2 = 6 \qquad (3.34)$$

As ℓ values are positive integers, the only choice of these values implied by this equation are 1, 1, 2. The first two irreps are one-dimensional and the third is two-dimensional. For Γ_1, we write

	E	$2C_3$	$3\sigma_v$
Γ_1	1	1	1

Components of Γ_2 stem from the orthogonality rule **III**. There results

	E	$2C_3$	$3\sigma_v$
Γ_2	1	1	-1

Note in particular that

$$\sum_R \chi_1(R)\chi_2(R) = (1) \times (1) + 2(1) \times (1) + 3(1) \times (-1) = 0$$

Here we have recalled rule **IV**: elements of the same class have the same character. As noted above, the third irrep is two-dimensional. The orthonormality rule **III** determines the characters of this irrep. We write

$$\sum_R \chi_1(R)\chi_3(R) = (1) \times (2) + 2(1) \times [\chi_3(C_3)] + 3(1) \times [\chi_3(\sigma_v)] = 0,$$

$$\sum_R \chi_2(R)\chi_3(R) = (1) \times (2) + 2(1) \times [\chi_3(C_3)] + 3(-1) \times [\chi_3(\sigma_v)] = 0$$

Solving these algebraic equations for $\chi_3(C_3), \chi_3(\sigma_v)$ gives

$$\chi_3(\sigma_v) = 0, \qquad \chi_3(C_3) = -1$$

These findings agree with the character values shown in table (3.27).

The principal manner in which rules **II** and **III** of the GOT come into play in a character table is shown schematically in Fig. 3.3.

3.5 Character Tables. Bases

A character table of a group is a display of the characters of elements of irreps of the group. As the characters of each member of a given class are equal, characters are listed according to classes of the group, such as shown in (3.23). The general form of a character table is shown in Table 3.1, again for the case of the C_{3v} group.

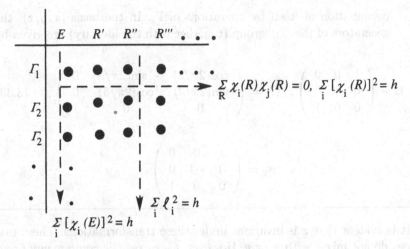

Figure 3.3. Three rules of the GOT related to the display of a character table. In these rules χ_i includes the class multiplicity factor. Wide dots represent character entries. The sum on R is over parallel rows.

Table 3.1

Character Table for the C_{3v} Group

C_{3v}	E	$2C_3$	$3\sigma_v$		
A_1	1	1	1	z	$x^2 + y^2,\ z^2$
A_2	1	1	-1	R_z	
E	2	-1	0	$(x, y)\quad (R_x, R_y)$	$(x^2 - y^2, xy)\quad (xz, yz)$
I		II		III	IV

Explanation of Domains

I. Names of irreps. All one-dimensional irreps are labeled either A or B. Irreps that are symmetric with respect to C_n rotations are labeled A, whereas those that are antisymmetric with respect to C_n rotations are labeled B. Two-dimensional irreps are labeled E. Three-dimensional irreps are labeled T and sometimes F.

II. Values of characters of group elements in the cited class.

III. This component of the character table represents *bases* of given irreps, and always includes the elements and combinations of the six symbols $x, y, z, R_x R_y, R_z$ where, for example, R_z represents rotation about the z axis and may be described by an arrow normal to the xy plane. To say that R_z is invariant under a given operation means that the direction of this arrow is invariant. A set of ℓ_i functions is a basis of an ℓ_i-dimensional irrep, Γ_i, if this set is transformed into a linear combination of itself by operations of Γ_i. In the basis (x, y, z), the generators of the C_{3v} group (together with the identity) are given by

$$
E = \begin{pmatrix} 1 & 0 & 0 \\ 0 & 1 & 0 \\ 0 & 0 & 1 \end{pmatrix}, \quad C_3 = \begin{pmatrix} \cos(2\pi/3) & -\sin(2\pi/3) & 0 \\ \sin(2\pi/3) & \cos(2\pi/3) & 0 \\ 0 & 0 & 1 \end{pmatrix}, \quad (3.35)
$$

$$
\sigma_v = \begin{pmatrix} 1 & 0 & 0 \\ 0 & -1 & 0 \\ 0 & 0 & 1 \end{pmatrix}.
$$

It is evident that z is invariant under these transformations. These matrices do not mix z with x or y. However, C_3 mixes the components (x, y) to give (x', y') so that (x, y) form a basis for the two-dimensional M matrices listed in (3.24) or, equivalently, the E representation. One also says, '(x, y) transforms according to E.' Furthermore, z is a basis for the one-dimensional Γ_1 group or, equivalently, the A_1 representation. One pictures R_z as a directed loop about the z axis in the xy plane so that E and C_3 elements of the irrep E have no effect on R_z but σ_v reverses it. Thus, R_z does not transform as the E irrep. Character elements of the A_2 irrep are $(1, 1, -1)$ and we see that R_z transforms as the A_2 irrep.

In column IV are found algebraic forms with invariance or parallel transformation properties, which serve as basis terms as well. For example, in the E irrep, the pair of functions (xz, yz) must have the same transformation properties as (x, y) since z is invariant under these transformations. Transforming $x^2 - y^2$ or xy by C_3 gives a linear combination of these forms, so that they too form a basis of the E irrep (Problem 3.11). The forms corresponding to the A_1 irrep are relevant to rotations in the plane for which $x^2 + y^2$ as well as z^2 are invariant.

A compilation of character tables for point groups is listed in Appendix A.

3.6   Representations of Cyclic Groups

In this section we apply preceding analysis to the cyclic C_n groups. We recall that every element of a cyclic group is in its own class. With rule **V** it follows that the group C_n has n irreps and with rule **I** (3.20) that all irreps are one-dimensional. We consider specifically the C_3 group with irreps $(\Gamma_1, \Gamma_2, \Gamma_3)$. The operation C_3 is conveniently written

$$
\varepsilon \equiv \exp(2\pi i/3), \quad i = 1, 2, 3.
$$

We list the following characters of the C_3 group and test for consistency with the GOT properties:

$$
\begin{array}{c|ccc}
C_3 & C_3 & C_3^2 & E \\
\hline
\Gamma_1 & \varepsilon & \varepsilon^2 & \varepsilon^3 \\
\Gamma_2 & \varepsilon^2 & \varepsilon^4 & \varepsilon^6 \\
\Gamma_3 & \varepsilon^3 & \varepsilon^6 & \varepsilon^9
\end{array}
\tag{3.36}
$$

Consider the property (3.26) [augmented in accord with (3.36) for complex operations]. We first test for orthogonality. Consider, for example,

$$
\sum_R \Gamma_2^* \Gamma_3 = (\varepsilon^2)^* \varepsilon^{2+1} + (\varepsilon^4)^* \varepsilon^{4+2} + (\varepsilon^6)^* \varepsilon^{6+3}
$$
$$
= \varepsilon + \varepsilon^2 + \varepsilon^3 = \sum_{n=1}^{3} \exp(2\pi i n/3) = 0.
\tag{3.37}
$$

In the complex plane, the right-hand side of (3.38) represents the sum of three uniformly oriented unit vectors which add to zero. The normalization implied by (3.22) is evidently satisfied. For example,

$$
\sum_R \Gamma_2^* \Gamma_2 = (\varepsilon^2)^* \varepsilon^2 + (\varepsilon^4)^* \varepsilon^4 + (\varepsilon^6)^* \varepsilon^6 = 1 + 1 + 1 = 3
\tag{3.38}
$$

We wish to cast the characters listed in (3.37) in more standard form. First, note that all elements in the third column and the third row are unity, and we may write

$$
\begin{array}{c|ccc}
C_3 & C_3 & C_3^2 & E \\
\hline
\Gamma_1 & \varepsilon & \varepsilon^2 & 1 \\
\Gamma_2 & \varepsilon^2 & \varepsilon^4 & 1 \\
\Gamma_3 & 1 & 1 & 1
\end{array}
\tag{3.39}
$$

Rearranging terms gives the more standard character table

$$
\begin{array}{c|ccc}
C_3 & E & C_3 & C_3^2 \\
\hline
A & 1 & 1 & 1 \\
E & 1 & \varepsilon & \varepsilon^* \\
E & 1 & \varepsilon^* & \varepsilon
\end{array}
\tag{3.40}
$$

Adding the two components of the E irrep gives the equivalent table

$$
\begin{array}{c|ccc}
C_3 & E & C_3 & C_3^2 \\
\hline
A & 1 & 1 & 1 \\
E & 2 & 2\cos 2\pi/3 & 2\cos 4\pi/3
\end{array}
\tag{3.41}
$$

(Here we have noted that $\cos 4\pi/3 = \cos[2\pi - 4\pi/3] = \cos 2\pi/3$.)

Let us demonstrate that characters of the E irrep listed above correspond to respective 2×2 matrix representations of these operations. These rotation matrices, including the identity matrix, are given by

$$
E = \begin{pmatrix} 1 & 0 \\ 0 & 1 \end{pmatrix}, \qquad
C_3 = \begin{pmatrix} \cos 2\pi/3 & -\sin 2\pi/3 \\ \sin 2\pi/3 & \cos 2\pi/3 \end{pmatrix},
$$

$$C_3^2 = \begin{pmatrix} \cos 4\pi/3 & -\sin 4\pi/3 \\ \sin 4\pi/3 & \cos 4\pi/3 \end{pmatrix}. \tag{3.42}$$

Traces of these matrices agree with characters listed in (3.41). Note also that the dimensions of the irreps listed in (3.41) agree with rule **I** (3.28).

As noted previously, there is only one group of prime order and any such group has no subgroups. However, as evidenced from the preceding discussion and rule **I** (3.20), a group of prime order may have a number of irreps.

Complex Groups

The representations of cyclic groups given above are an example of complex groups. For this case, the diagonal component of rule **III** (3.30) is written

$$\sum_R \chi_i(R)\chi_i(R)^* = h. \tag{3.43}$$

This relation may be taken as a criterion for irreducibility of a given representation $\Gamma_i(R)$.[1] As the components of Γ_i^* are complex conjugates of Γ_i with (3.43) it follows that if Γ_i is an irrep, so is Γ_i^*. This conclusion is relevant to the complex representation of a cyclic group described above, so that the character table (3.41) implies that there is another irrep which is a complex conjugate of the given irrep. As discussed in the following chpter, dimensionality of irreps is related to degeneracy of quantum states. Dimensionality of irreps for this purpose is best discovered by rewriting the character table in terms of real values, such as given in (3.41). However, it should be borne in mind that the real character table (3.41) is not a consistent group theoretic character table. For example, table (3.41) does not satisfy the basic irrep rules given above whereas the complex representation (3.40) does satisfy these rules.

The exponential representation has specific relevance to cyclic groups of odd order. In this event with the GOT, the character table cannot be written in terms of 1 or −1 elements only, but may be written in terms of these complex exponential forms. Thus, for example, the character table for C_3 (3.40) is consistently written in terms of $1, \varepsilon$ and ε^* terms.

---

[1]L. Jansen and M. Boon, *Theory of Finite Groups, Application in Physics* (see Bibliography).

Summary of Topics for Chapter 3

1. Matrix representation of operators.

2. Character of a matrix.

3. Basic matrix definitions.

4. Irreduciable representations (irreps).

5. Great orthogonality theorem (GOT).

6. Matrix representation of groups.

7. Similarity transformation.

8. Relations between character, order, class and irrep dimensions of a given group.

9. Character tables.

10. Representation of cyclic groups.

11. Complex groups.

Problems

3.1 Establish rule **IV** of Section 3.4.

3.2 (a) Show that the eigenvalues of an Hermitian matrix are real. (b) Show that the eigenvectors of an Hermitian matrix are orthogonal.

3.3 Show that the irreps of a cyclic group are all one-dimensional.

3.4 Consider the character table for the C_{6v} group (Appendix A). Employing the orthonormality rule **III**, Section 3.4, establish the four zero character entries of the E_1 and E_2 irreps.

3.5 Establish rule **III**, Section 3.4, for normalization (i.e., the case $i = j$).
Answer
Employing the GOT we write

$$\sum_R \Gamma_i(R)_{mm} \Gamma_i(R)_{m'm'} = \frac{h}{\ell_i} \delta_{mm'} \delta_{mm'} \tag{P1}$$

Now sum the left side over m, m'.
There results

$$\sum_m \sum_{m'} \sum_R \Gamma_i(R)_{mm} \Gamma_i(R)_{m'm'} = \sum_R \left\{ \sum_m \Gamma_i(R)_{mm} \sum_{m'} \Gamma_i(R)_{m'm'} \right\}$$

$$= \sum_R \chi_i(R) \chi_i(R) = \frac{h}{\ell_i} \sum_m \sum_{m'} \delta_{mm'} \delta_{mm'} = \frac{h}{\ell_i} \sum_m \delta_{mm} = \frac{h}{\ell_i} \ell_i = h$$

where (P1) was employed in the last sequence. Equating the first and last terms of the last line returns rule **III**.

3.6 (a) How many irreps does the C_5 cyclic group have? (b) List the rotation matrices corresponding to the C_5 group and their respective characters. (c) Repeat the construction of (3.40) for the case of the C_5 cyclic group and establish orthogonality of two irreps of your table.

3.7 Show that every group of prime order is Abelian.
Answer
Every group of prime order is cyclic and every cyclic group is Abelian.

3.8 (a) What are the possible subgroups of the group of order 17? (b) What are the possible dimensions of the irreps of this group?

3.9 Construct the matrix representation of the D_2 group $[E, C_2(z), C_2'(x), C_2'(y)]$.

3.10 Show that $\Gamma_1 = \{1, -1, -1, 1\}$ is homomorphic to the C_{2v} group $\{E, C_2, \sigma_v, \sigma_v'\}$, i.e., show that these two groups have parallel group tables.

3.11 (a) Applying the 2×2 matrix, C_n, to the column vector (x, y), obtain the form that the terms $(x^2 - y^2, xy)$ go to under this transformation. Leave your answer in terms of trigonometric functions. Write the answer in the following form.

$$\begin{pmatrix} x^2 - y^2 \\ xy \end{pmatrix} = D \begin{pmatrix} x'^2 - y'^2 \\ x'y' \end{pmatrix}$$

where D is the implied matrix. (b) Show that for $\theta = 2\pi/3, \chi(D) = -1$.

3.12 How many irreps does the C_n have? Why? What are the dimensions of these irreps? (b) Write down a character table of an irrep of the C_6 group in terms of 1 and -1 only. (c) Show that rules **II** and **III** (Section 3.4) are satisfied for this irrep. (d) Consider the C_n group for n odd. Does a character table for this group exist in terms of 1 and -1 only? Explain your answer.

Answer (partial)

(a) Every element of a cyclic group is in a class by itself. The number of irreps of a group is equal to the number of classes of the group. It follows that the C_n group has n irreps. These irreps are one-dimensional.

3.13 Show that a real unitary matrix is orthogonal.

4

Quantum Mechanics, the Full Rotation Group, and Young Diagrams

4.1 Application to Quantum Mechanics

The time-independent Schrödinger equation is given by

$$H\psi = E\psi \qquad (4.1)$$

where H represents the Hamiltonian operator. For a point particle in a potential field $V(\mathbf{r})$, H is given by

$$H = \frac{p^2}{2m} + V(\mathbf{r}) \qquad (4.2)$$

where $\mathbf{p}$ is the momentum operator

$$\mathbf{p} = -i\hbar\nabla \qquad (4.3)$$

Note that (4.1) is an eigenvalue equation in which the eigenfunction ψ corresponds to the eigenvalue E. We wish to consider cases for which the system at hand has symmetries. In this case, eigenenergies will be degenerate. An eigenenergy, E_q, is q-fold degenerate if there exist q independent eigenfunctions corresponding to E_q. Consider the case of a particle confined to the interior of a square in the plane with perfectly reflecting walls. Suppose ψ_q is an eigenfunction of the related Hamiltonian with energy E_q. Rotation of coordinates about the center of the square through $\pi/2$ produces a new eigenfunction which evidently also corresponds to the eigenenergy E_q. Continuing this process produces four eigenfunctions which correspond to the eigenenergy E_q. This degeneracy is evidently related to the C_4 symmetry of the square. A very useful application of group theory enables one to ascertain degeneracies of a given system in terms of properties of the irreps the group of symmetries of the system.

Let R denote a symmetry operation on a given system (e.g., an atom or a molecule) with the Hamiltonian H. It follows that H and R commute,

$$RH = HR \qquad (4.4)$$

We wish to establish the following property of degenerate functions: Let $\{\psi_i\}, i = 1, 2, \ldots, q$, be eigenfunctions corresponding to the eigenenergy E_q. Then any linear combination of $\{\psi_i\}$ is also an eigenfunction of H corresponding to the eigenenergy E_q,

$$H \sum_{n=1}^{q} a_n \psi_n = \sum_{n=1}^{q} a_n H \psi_n = E_q \sum_{n=1}^{q} a_n \psi_n \qquad (4.5)$$

which establishes the property. In the following it is assumed that eigenfunctions of H comprise an orthonormal sequence (Problem 4.1), that is,

$$\int \psi_i^* \psi_j \, d\mathbf{r} = \delta_{ij} \qquad (4.6)$$

where the integral extends over the accessible domain of the system.

Eigenfunction Basis

Non-Degenerate Eigenstates

We wish to establish the following important property: Eigenfunctions of a system with a group of symmetries G are bases of the irreps of G. We first establish this property for the non-degenerate eigenfunctions of the system. Let ψ_i denote one such non-degenerate eigenfunction. With the commutation property (4.4), we write

$$HR\psi_j = RH\psi_j = RE_j\psi_j = E_j R\psi_j \qquad (4.7)$$

where, again, R denotes any of the symmetry operations of G. Comparing the first and last terms in the latter equalities indicates that $R\psi_j$ is an eigenfunction of H. With (4.6) one obtains (with R real)

$$\int \psi_i^* \psi_j \, d\mathbf{r} = \int R\psi_i^* R\psi_j \, d\mathbf{r} = \delta_{ij} \qquad (4.8a)$$

so that

$$R\psi_j = \pm 1 \psi_j \qquad (4.8b)$$

Eigenvalues of R are ± 1. Applying each of the R operations to a non-degenerate eigenfunction of the system produces a representation of the group of symmetries of the given system, with each matrix, $\Gamma_i(R) = \pm 1$. As these matrices are one-dimensional they are irreducible.

Degenerate Eigenstates

Let the system at hand have q-fold degeneracy. In this event (4.7) generalizes to

$$HR\psi_{jn} = E_j R\psi_{jn} \tag{4.9a}$$

$$n = 1, 2, \ldots, q \tag{4.9b}$$

As $R\psi_{jn}$ may be any linear combination of ψ_{jn} functions we write

$$R\psi_{jn} = \sum_{k=1}^{q} r_{kn}\psi_{jk} \tag{4.10a}$$

Let S denote another symmetry operation of the system. We write

$$S\psi_{jv} = \sum_{p=1}^{q} s_{pv}\psi_{jp} \tag{4.10b}$$

Because S and R are members of G so is $T = SR$, and we write

$$T\psi_{jp} = \sum_{v=1}^{q} t_{vp}\psi_{jv} \tag{4.10c}$$

Applying separate effects of S and R gives

$$SR\psi_{jp} = S\sum_{v=1}^{q} r_{vp}\psi_{jv} = \sum_{v=1}^{q}\sum_{i=1}^{q} s_{iv}r_{vp}\psi_{ji} \tag{4.10d}$$

Comparing the latter expression with (4.10c) we find

$$t_{ip} = \sum_{v=1}^{q} s_{iv}r_{vp} \tag{4.11a}$$

or, equivalently, in matrix notation

$$T = SR \tag{4.11b}$$

This is the expression of the element of a q-dimensional matrix T in terms of the product of two other q-dimensional matrices, R and S. It follows that matrices corresponding to the transformation of q eigenfunctions related to a q-fold degenerate eigenenergy are a q-dimensional irrep of the group. If this representation were reducible, the set of q eigenfunctions could be divided into subsets corresponding to different eigenenergies, thus contradicting the assumption that the sequence described in (4.9) has a common eigenenergy.

Irreps and Degeneracy

In this manner we uncover the very useful rules:

A. Orders of degeneracy of a system with a Hamiltonian with symmetries described by the group G are given by the dimensions of the irreps of G.

B. The q degenerate eigenstates of the Hamiltonian comprise the basis of a q-dimensional irrep of the group G. Non-degenerate states comprise one-dimensional irreps of the group G.

In application of rule A, one need only consult the character table of the given group. As discussed previously, the first column of a character table gives the dimensions of irreps of the group, which in turn indicate degeneracies of quantum states of the system.

Consider the quantum dynamics of a particle confined to the interior of a cube, whose symmetries are described by the O_h group (2.13). Consulting the character table for this group indicates that the maximum degeneracy of a state is threefold. Eigenenergies for this configuration are proportional to $(n_1^2 + n_2^2 + n_3^2)$ where n_1, n_2, n_3 are positive integers. If $n_1 \neq n_2 \neq n_3$, then one expects that this state is sixfold degenerate $(3! = 6)$. However, the reader may convince himself that these degenerate states are not linearly independent. That is, any of these six states may be written as a linear combination of three linearly independent functions and, consistent with group theory, the degeneracy of this state is threefold. Another example is that of the quantum dynamics of a particle confined to the interior of an equilateral triangle prism, with related C_{3v} symmetry. The identity column of the character table for this group is shown below. The one-dimensional irreps in this table may be identified with the ground state. The table indicates that all other states are twofold degenerate:

C_{3v}	E
A_1	1
A_2	1
E	2

Note that the character table of any C_n group indicates that all irreps are one-dimensional. This feature stems from the following properties: all elements of a cyclic group are in a class by themselves, rule **V** (Chapter 3) the number of irreps of a group equals the number of classes of the group, and rule **I** (Chapter 3) the sum of the squares of the dimensions of the irreps of a group equals the order of the group.

The relation between order of degeneracy and dimension of irreps is summarized below:

Degeneracy	Dimension of Irrep
Non-degenerate	$\ell_i = 1$
q_i-fold degenerate	$\ell_i = q_i$.

In general, s-fold degenerate eigenfunctions, of a system Hamiltonian that commutes with the symmetry operations of the system, comprise the basis functions of an s-dimensional irrep of that symmetry group.

Note also that a quantum system with no degeneracies is void of symmetries and, conversely, a quantum system with no symmetries is non-degenerate. The symmetry group describing an asymmetric system has one element, the identity. Perturbations that break symmetries remove degeneracies.

4.2 Full Rotation Group $O(3)$

Representations of the Full Rotation Group; Euler Angles

Proper rotation about a fixed point taken as the origin of an orthogonal x, y, z axes may be written

$$R(\alpha, \beta, \gamma) = R_x(\alpha) R_y(\beta) R_z(\gamma) \qquad (4.12a)$$

$$\mathbf{r}' = R(\alpha, \beta, \gamma)\mathbf{r} \qquad (4.12b)$$

where $\mathbf{r}'$ is the $\mathbf{r}$ vector in the rotated frame,

$$R_z(\gamma) = \begin{pmatrix} \cos\gamma & \sin\gamma & 0 \\ -\sin\gamma & \cos\gamma & 0 \\ 0 & 0 & 1 \end{pmatrix}, \dots \qquad (4.12c)$$

and (α, β, γ) are the Euler angles.

Euler angles are defined in the following manner. $R(\gamma)$ is a rotation through γ about the z axis. New axes are labeled $x_1, y_1, z_1 = z$. $R(\beta)$ is a rotation through β about the y_1 axis. New axes are labeled $x_2, y_2 = y_1, z_2$. $R(\alpha)$ is a rotation through α about the x_2 axis. New axes are labeled $x_3 = x_2, y_3, z_3$.

The full rotation group is a continuous point group containing symmetry operations on a sphere. It is called $O(3)$ in reference to the property that it contains all real orthogonal transformations about a fixed point in 3-space. Such mappings consist of all transformations with real coefficients which leave the unit sphere

$$x^2 + y^2 + z^2 = 1$$

invariant. The group $O(3)$ contains inversion and proper rotations. The group which contains proper rotations only is labeled the $O(3)^+$ group. Recalling the inversion group, $C_i = \{E, i\}$, the full rotation group $O(3)$ may be written as the direct product

$$O(3) = O(3)^+ \otimes C_i \qquad (4.13)$$

The infinite continuous group $O(3)^+$ is composed of all possible $R(\alpha, \beta, \gamma)$ matrices.

Irrep Basis Property

Consider that the symmetries of a Hamiltonian comprise the group G. Let s eigenstates of H have the following property. The action of any element of G on these s eigenstates transforms them among themselves. Equivalently one may say that these eigenstates span a subspace, S, of Hilbert space which is invariant to operations of G. The addition or subtraction of any other eigenstate of H to this set of s functions breaks this invariance property. The dimension, s, of this subspace is irreducible in the sense that actions of G on elements of S do not carry elements outside of S, so that S is an invariant subspace. One concludes that the s eigenstates are basis functions for irreps of G.

Basis Functions

To obtain basis functions for the group $O(3)^+$ we examine Laplace's equation

$$\nabla^2 \psi(x, y, z) = 0 \tag{4.14}$$

Let R denote an element of the $O(3)^+$ group. As ∇^2 is a rotationally symmetric operator, it follows that $R\nabla^2 - \nabla^2 R$, so that if ψ is a solution to (4.14), so is $R\psi$. Working in spherical coordinates (θ and ϕ are polar and azimuthal angles, respectively), solutions to (4.14) are given by

$$\psi = r^\ell Y_{\ell m}(\theta, \phi) \tag{4.15a}$$

where $Y_{\ell m}(\theta, \phi)$ are spherical harmonics defined by the relation

$$Y_{\ell m}(\theta, \phi) = e^{im\phi} P_{\ell m}(\theta) \tag{4.15b}$$

where ℓ is a non-negative integer and $P_{\ell m}(\theta)$ are associated Legendre functions. Spherical harmonics are such that for a given value of ℓ, there are $2\ell + 1$ linearly independent solutions of (4.14) corresponding to m values

$$m = -\ell, \ldots, 0, \ldots, \ell \tag{4.15c}$$

These $2\ell + 1$ functions transform only among themselves under any operation of $O(3)^+$, so that

$$R(\alpha, \beta, \gamma) Y_{\ell m}(\theta, \phi) = \sum_{m'=-\ell}^{\ell} = D^\ell(\alpha, \beta, \gamma)_{m'm} Y_{\ell m'}(\theta, \psi) \tag{4.16}$$

With the preceding irrep basis property, it follows that the set of functions $\{Y_{\ell m}(\theta, \phi)\}$ at any ℓ, forms a $(2\ell + 1)$ odd-dimensional basis of an irrep of $O(3)^+$. The coefficients $D^\ell(\alpha, \beta, \gamma)_{m'm}$ in (4.16) represent elements (i.e., matrices) of these irreps. (D corresponds to the German, "darstellung," i.e., representation.)

The characters of the D^ℓ elements may be obtained as follows. Consider a rotation about the z axis through γ,

$$R(0,0,\gamma)Y_{\ell m}(\theta,\phi) = Y_{\ell m}(\theta,\phi-\gamma) = e^{-im\gamma}Y_{\ell m}(\theta,\phi) \qquad (4.17)$$

Comparing this relation with (4.16) indicates that

$$D^\ell(0,0,\gamma)_{m'm} = e^{im\gamma}\delta_{mm'} \qquad (4.18)$$

which is a diagonal matrix. It follows that the character of the D^ℓ matrix corresponding to rotation through γ is given by

$$\chi^\ell(\gamma) = \sum_{m=-\ell}^{\ell} e^{-im\gamma} = \frac{\sin[(\ell+\frac{1}{2})\gamma]}{\sin(\gamma/2)} \qquad (4.19)$$

This is the character of D^ℓ for rotation through γ about any axis.

We recall that elements of the same character belong to the same class. From the preceding result it follows that for the $O(3)^+$ group, rotations through a given angle about any axis belong to the same class. Thus there are an infinite number of classes of $O(3)^+$ which infers that there are an infinite number of irreps for this group.

There are no other odd-dimensional irreps of $O(3)^+$ other than D^ℓ. If another irrep Γ' exists, its character is orthogonal to (4.19) [recall (3.21)] for all ℓ values or, equivalently, it is orthogonal to the difference between two successive characters

$$\chi^{\ell-1} - \chi^\ell = 2\cos\ell\gamma, \quad 0 \le \gamma \le \pi \qquad (4.20)$$

It follows that the character of Γ' is orthogonal to a cosine Fourier series which forms a complete set in the range $(0, 2\pi)$ so there can be no other independent representations. One may conclude that D^ℓ are the only irreps of $O(3)^+$ and that the spherical harmonics are basis functions for these odd-dimensional irreps of $O(3)^+$. Note that although $O(3)^+$ is a continuous group its irreps comprise a countable set.

4.3 $SU(2)$

Basic Properties

Having discovered the odd-dimensional irreps of $O(3)^+$, we now consider irreps of even dimension. The symbol $SU(2)$ denotes 'special unitary,' two-dimensional. Let U be an element of $SU(2)$. Then

$$U^\dagger U = I \text{ and } |U| = 1 \qquad (4.21)$$

where I represents the identity matrix and, again, $|U|$ denotes the determinant of U. The right equality in (4.21) is the 'special' quality of these elements. The generic form of an element of $SU(2)$ is given by

$$U = \begin{pmatrix} a & b \\ -b^* & a^* \end{pmatrix} \tag{4.22}$$

for which

$$|U| = |a|^2 + |b|^2 = 1$$

With

$$U^\dagger = \begin{pmatrix} a^* & -b \\ b^* & a \end{pmatrix} = U^{-1}$$

we see that the left equality of (4.21) is satisfied as well.

Relation to $O(3)^+$

We wish to establish a relation between $SU(2)$ and $O(3)^+$. Consider the 2×2 matrices

$$A = \begin{pmatrix} -z & x+iy \\ x-iy & z \end{pmatrix}, \quad A' = \begin{pmatrix} -z' & x'+iy' \\ x'-iy' & z' \end{pmatrix} \tag{4.23a}$$

where $A' \equiv A(\mathbf{r}')$ and $\mathbf{r} = (x, y, z)$. The matrix A' is related to A through the similarity transformation

$$A' = UAU^{-1} \tag{4.23b}$$

where U is given by (4.22). The transformation (4.23b) implies the following relations:

$$x' = \frac{1}{2}[a^2 + a^{*2} - b^2 - b^{*2}]x + \frac{1}{2}i[a^2 - a^{*2} + b^2 - b^{*2}]y + (ab + a^*b^*)z \tag{4.23c}$$

$$y' = \frac{1}{2}i[a^{*2} - a^2 + b^2 - b^{*2}]x + \frac{1}{2}[a^2 + a^{*2} + b^2 + b^{*2}]y + i(a^*b^* - ab)z \tag{4.23d}$$

$$z' = -(a^*b + ab^*)x + i(a^*b - ab^*)y + (aa^* - bb^*)z \tag{4.23e}$$

Thus, we may write

$$\mathbf{r}' = R(a, b)\mathbf{r} \tag{4.24a}$$

One finds that

$$R\tilde{R} = I, \quad |R| = 1 \tag{4.24b}$$

It follows that R represents pure rotation and may be related to Euler angles. Consider the case

$$a = e^{-i\alpha/2}, \quad b = 0 \tag{4.25a}$$

With (4.22) we obtain

$$U(\alpha) = \begin{pmatrix} e^{-i\alpha/2} & 0 \\ 0 & e^{i\alpha/2} \end{pmatrix} \tag{4.25b}$$

which, with (4.24a), gives

$$R(\alpha) = \begin{pmatrix} \cos\alpha & \sin\alpha & 0 \\ -\sin\alpha & \cos\alpha & 0 \\ 0 & 0 & 1 \end{pmatrix} \qquad (4.25c)$$

In the second example we choose

$$a = -e^{-\alpha/2}, \quad b = 0 \qquad (4.26a)$$

which gives

$$-U(\alpha) = \begin{pmatrix} -e^{-i\alpha/2} & 0 \\ -0 & -e^{i\alpha/2} \end{pmatrix} \qquad (4.26b)$$

This form returns the same $R(\alpha)$ as given by (4.25c). We may conclude that $U(\alpha)$ and $-U(\alpha)$ in $SU(2)$ correspond to the same rotation matrix $R(\alpha)$ in $O(3)^+$. Another example is given by

$$a = a^* = \cos\frac{\beta}{2}, \quad b = b^* = -\sin\frac{\beta}{2} \qquad (4.27a)$$

which corresponds to

$$U(\alpha) = \begin{pmatrix} \cos\frac{\beta}{2} & -\sin\frac{\beta}{2} \\ \sin\frac{\beta}{2} & \cos\frac{\beta}{2} \end{pmatrix} \qquad (4.27b)$$

and

$$R(\beta) = \begin{pmatrix} \cos\beta & 0 & -\sin\beta \\ 0 & 1 & 0 \\ \sin\beta & 0 & \cos\beta \end{pmatrix} \qquad (4.27c)$$

which relates to rotation through β about the y axis which again may be described by $\pm U(\beta)$.

The complete Euler angle transformation is generated by the relation

$$a = [\exp -i(\alpha + \gamma)/2]\cos\frac{\beta}{2}$$
$$b = -[\exp -i(\alpha - \gamma)/2]\sin\frac{\beta}{2} \qquad (4.27d)$$

which gives the full rotation matrix $R(\alpha, \beta, \gamma)$ (Problem 4.7). As there is a two-to-one relation between the elements of $SU(2)$ and $O(3)^+$ one may say that $SU(2)$ is homomorphic to $O(3)^+$.

Irreps of SU(2)

Let the column vector on which U matrices of $SU(2)$ have components (u, v) so

$$\begin{pmatrix} u' \\ v' \end{pmatrix} = U(a, b) \begin{pmatrix} u \\ v \end{pmatrix} \qquad (4.28a)$$

which, with the basic form of U (4.22), gives

$$u' = au + bv$$
$$v' = -b^*u + a^*v$$

(4.28b)

Consider the $n + 1$ polynomials in u and v,

$$u^n, u^{n-1}v, \ldots, uv^{n-1}, v^n$$

These $n+1$ polynomials transform among themselves under the operations $U(a, b)$ of $SU(2)$. It follows that these polynomials form an $(n + 1)$-dimensional basis for representations of $SU(2)$. Setting $n = 2j$, the preceding list of polynomials becomes

$$u^{2j}, u^{2j-1}v, \ldots, uv^{2j-1}, v^{2j}$$

As n is an integer, j is an integer or a half odd integer. For example, $j = 1/2$ gives the polynomials u, v; $j = 1$ gives u^2, uv, v^2; and $j = 3/2$ gives u^3, u^2v, uv^2, v^3.

These basis functions are all in the form $u^{j+1}v^{j-1}$, where $-j \leq m \leq j$ in unit steps, and are written

$$f_j^m(u, v) = \frac{u^{j+m}v^{j-m}}{[(j+1)!(j-1)!]^{1/2}}, \quad m = j, \ldots, j$$

(4.20)

where j may be integer or half-odd integer. The denominator ensures that the corresponding representation is unitary. Consider the operation

$$Uf_j^m(u, v) = f_j^m(a^*u - bv, b^*u + av)$$

$$= \frac{(a^*u - bv)^{j+m}(b^*u + av)^{j-m}}{[(j+1)!(j-1)!]^{1/2}}$$

(4.30)

where the effect of the matrix U on polynomial forms in uv is given by (4.28). Expanding the right side of this relation with the binomial theorem and expressing u and v in terms of f_m^j functions gives

$$Uf_m^j = \sum_{m'} U_{m'm}^{(j)}(a, b)f_m^j$$

(4.31a)

where

$$U_{m'm}^{(j)}(a, b) = \sum_p \frac{(-1)^p[(j+m)!(j-m)!(j+m')!]^{\frac{1}{2}}}{p!(j-m'-p)!(j+m-p)!(p+m'-m)!}$$

$$\times a^{j-m'-p}a^{*(j+m-p)}b^pb^{*(p+m'-m)}$$

(4.31b)

The summation runs over all integer p for which the denominator is non-zero (recall $0! = 1$).

Special Values

We wish to show that the $U^{(j)}_{m'm}(a,b)$ matrices comprise an irreducible representation of $SU(2)$. Toward these ends we consider two special values of these matrices. Consider first the case $b = 0$, for which all terms in $U^{(j)}_{m'm}(a,b)$ with $p \neq 0$ are zero and when $p = 0$ all terms with $m - m' \neq 0$ are zero. The only non-vanishing term corresponds to $p = 0$ and $m = m'$,

$$U^{(j)}_{m'm}(a,0) = \delta_{mm'} a^{j-m} a^{*(j-m)} \tag{4.32a}$$

The value $a = e^{-i\alpha/2}$ gives the diagonal matrix

$$U^{(j)}_{m'm}(e^{i\alpha/2}, 0) = \delta_{mm'} e^{-im\alpha} \tag{4.32b}$$

In the second case we set $j = m'$. The term in the denominator of (4.31b), $(j - m' - p)!$, remains finite only if $p = 0$. There results

$$U^{(j)}_{m'm}(a,b) = \sqrt{\frac{(2j)!}{(j+m)!(j-m)!}} a^{*(j+m)} b^{*(j-m)} \tag{4.32c}$$

Schur's Lemma Revisited

With the special values (4.32b,c) at hand we return to Schur's lemma (Section 3.4) which, we recall, states that if the only matrix that commutes with all elements of a given representation is a constant matrix (a constant times the identity matrix), then the representation is irreducible. Assume that a matrix M exists which commutes with $U^{(j)}(a,b)$,

$$MU^{(j)}(a,b) = U^{(j)}(a,b)M \tag{4.33a}$$

or, equivalently,

$$\sum_s M_{m's} U^{(j)}_{sm}(a,b) = \sum_s U^{(j)}_{m's}(a,b) M_{sm} \tag{4.33b}$$

As this relation holds for all $U^{(i)}(a,b)$, it holds for the diagonal form (4.32b) which gives

$$\sum_s M_{m's} \delta_{sm} e^{im\alpha} = \sum_s M_{sm} \delta_{sm} e^{im\alpha} M_{sm} \tag{4.33c}$$

Completing the summation one obtains

$$M_{m'm} e^{im\alpha} = e^{im'\alpha} M_{m'm} \tag{4.33d}$$

It follows that $M_{m'm} = 0$ unless $m = m'$, so that $M_{m'm}$ is a diagonal matrix. Thus (4.33b) reduces to

$$M_{m'm'} U^{(j)}_{m'm}(a,b) = U^{(j)}_{m'm'}(a,b) M_{mm} \tag{4.33e}$$

If, in addition, $j = m$,

$$M_{jj} U^{(j)}_{jm}(a,b) = U^{(j)}_{jm}(a,b) M_{mm} \tag{4.33f}$$

The matrix $U_{jm}^{(j)}(a,b)$ is given by (4.32c) which in general is not zero. It follows that $M_{jj} = M_{mm}$ for all m, which means that M is a diagonal matrix with equal diagonal elements. Thus M is the multiple of the unit matrix, and we find that the only matrix which commutes with the matrices $U^{(j)}(a,b)$ is a constant matrix. According to Shur's lemma, it follows that the matrices $U^{(j)}(a,b)$ comprise an irreducible representation of $SU(2)$.

4.4 Irreps of $O(3)^+$ and Coupled Angular Momentum States

As found above, the $O(3)^+$ and $SU(2)$ groups are homomorphic. This property provides a simple means of obtaining the irreps of $O(3)^+$. That is, one merely need replace the Euler equivalents for (a,b) given by (4.27a) into the $U_{m'm}^{(j)}(a,b)$ matrix elements of the irrep of $SU(2)$ to obtain an irrep of $O(3)^+$. An appropriately symmetrized form is constructed by multiplying the resulting expression by the unitary matrix $\delta_{m'm}(-1)^m$. There results

$$D_{m'm}^{(j)}(\alpha,\beta,\gamma) = \sum_p \frac{(-1)^p[(j+m)!(j-m)!(j+m')!]^{\frac{1}{2}}}{p!(j-m'-p)!(j+m-p)!(p+m'-m)!}$$
$$\times e^{im'\alpha}e^{im\gamma}\left(\cos\frac{\beta}{2}\right)^{(2j+m-m'-2p)}\left(\sin\frac{\beta}{2}\right)^{(2p+m'-m)}$$

$$(4.34)$$

At a given value of j, these are the matrix elements of an irreducible representation of $O(3)^+$.

Rotation Groups and Angular Momenta

In quantum mechanics, j values (integer or half-odd integer) are associated with the spin and orbital components of total angular momentum, $\mathbf{J}$, whereas ℓ values of the spherical harmonics are associated only with orbital angular momentum, $\mathbf{L}$. One writes

$$\mathbf{J} = \mathbf{L} + \mathbf{S} \qquad (4.35a)$$

where $\mathbf{S}$ denotes spin angular momentum. Angular momentum eigenvalues are of the form

$$J^2 = \hbar^2 j(j+1)$$
$$(4.35b)$$
$$m_j = -j, \ldots, j$$

The operator J^2 satisfies the relation

$$J^2 = L^2 + S^2 + 2\mathbf{L}\cdot\mathbf{S} \qquad (4.36a)$$

Eigenfunctions of orbital angular momentum are the spherical harmonics and obey the eigenvalue equation

$$L^2 Y_{\ell m}(\theta, \phi) = \hbar^2 \ell(\ell + 1) Y_{\ell m}(\theta, \phi)$$

$$L_z Y_{\ell m}(\theta, \phi) = \hbar m Y_{\ell m}(\theta, \phi)$$

(4.36b)

Eigennumbers (ℓ, m) are integers and satisfy parallel relations to (4.36a) [see also (4.15c)].

Addition of Angular Momentum

Let system No. 1 have total angular momentum $\mathbf{J}_1$ and let system No. 2 have total angular momentum $\mathbf{J}_1$. Eigenproperties of these operators are given by

$$J_1^2 |j_1, m_{j_1}\rangle = \hbar^2 j_1(j_1 + 1)|j_1, m_{j_1}\rangle, \qquad J_{1z}|j_1, m_{j_1}\rangle = \hbar m_{j_1}|j_1, m_{j_1}\rangle$$

$$J_2^2 |j_2, m_{j_2}\rangle = h^2 j_2(j_{21} + 1)|j_2, m_{j_2}\rangle, \qquad J_{2z}|j_2, m_{j_2}\rangle = \hbar m_{j_2}|j_2, m_{j_2}\rangle$$

(4.37a)

The Dirac 'ket' vectors in these equations are such that, say, projection of the ket vector, $|j, m\rangle$, onto the coordinate space 'bra' vector, $\langle \mathbf{r}|$, for integer $j = \ell$, gives the spherical harmonic eigenstate, $\langle \mathbf{r}|\ell, m\rangle = Y_{\ell m}(\theta, \phi)$. Consider the total angular momentum of the two systems, Nos. 1 and 2,

$$\mathbf{J} = \mathbf{J}_1 + \mathbf{J}_2 \tag{4.37b}$$

Eigenvalues of this operator are given by

$$j = |j_1 - j_2|, \ldots, |j_1 + j_2| \tag{4.37c}$$

This sequence of values runs in integer steps. Thus, for example, consider the case $j_1 = 3/2$ and $j_2 = 2$. Then total j numbers are given by $j = 1/2, 3/2, 5/2, 7/2$.

The component operators J_x, J_y, J_z obey the commutator relations

$$[J_x, J_y] = i\hbar J_z, \quad [J_y, J_z] = i\hbar J_x, \quad [J_z, J_x] = i\hbar J_y \tag{4.37d}$$

Spin eigenfunctions are column vectors. For $j = s = 1/2$, $m_s = -1/2, 1/2$, eigenfunctions are two-component column vectors called *spinors*. As remarked above, eigenfunctions corresponding to half-odd integer eigenvalues are disjoint from those of orbital angular momentum. Orbital angular momentum refers to orbits in real space spanned by three unit vectors. Spin(1/2) eigenstates exist in an abstract spin space spanned by the two column vectors $(1, 0), (0, 1)$. Every *spinor*, (u, v), in this space is a linear combination of these two unit vectors,

$$\begin{pmatrix} u \\ v \end{pmatrix} = u \begin{pmatrix} 1 \\ 0 \end{pmatrix} + v \begin{pmatrix} 0 \\ 1 \end{pmatrix} \tag{4.38a}$$

where, in general, u and v are complex numbers. This spin space is the space on which $U(a, b)$ matrices of $SU(2)$ operate [cf., (4.28a)]. In a representation in which S^2 and S_z are diagonal, spin operators are given by the matrices

$$S^2 = \frac{3}{4}\hbar^2 \begin{pmatrix} 1 & 0 \\ 0 & 1 \end{pmatrix}, \quad S_z = \frac{1}{2}\hbar \begin{pmatrix} 1 & 0 \\ 0 & -1 \end{pmatrix} \tag{4.38b}$$

As these operators commute they share eigenvectors (here labeled ξ) given by

$$\xi = \begin{pmatrix} 1 \\ 0 \end{pmatrix}, \begin{pmatrix} 0 \\ 1 \end{pmatrix} \tag{4.38c}$$

See Problem 4.13.

Representation for Infinitesimal Rotations

Let R denote a rotation operator. Then with (4.16) the transformation of the basis $\{\Psi_m\}$ is given by

$$R\psi_m = \sum_{m'} \psi_{m'} \mathbf{D}_{m'm}(R) \tag{4.39a}$$

Again, the expansion matrices $\mathbf{D}_{m'm}$ represent elements of irreps of the rotation group of the R rotations. When R is an infinitesimal rotation about the z axis, one writes

$$\mathbf{D}_{mm'}[R_\varepsilon(z)] = \delta_{mm'} - i\varepsilon(J_z)_{mm'} \tag{4.39b}$$

where component angular matrices obey the commutator relation (4.38d) and ε is an infinitesimal. For finite rotation, the relation (4.39b) generalizes to

$$\mathbf{D}^j_{mm'}[R_\theta(z)] = [\exp(i\theta J_z)]_{mm'} \tag{4.39c}$$

The j superscript relates to the eigenvalue of J^2 (4.37a). The characters of the $\mathbf{D}^j$ matrices are given by (4.19). However, summation is now over m_j numbers.

We wish to find the direct product of two $\mathbf{D}^j$ matrices, $\mathbf{D}^{j_1} \otimes \mathbf{D}^{j_2}$, relevant to the irreps of the rotation operator corresponding to the coupling of two angular momentum states (j_1, j_2). To these ends we write $\mathbf{D}^{j_1} \otimes \mathbf{D}^{j_2} = \sum b(j)\mathbf{D}^j$ and calculate $b(j)$. The sum is over $j = 0, 1, \ldots, \infty$. Working with characters simplifies the calculation. Thus, the character of the preceding direct product is given by

$$\chi^{j_1}(\gamma_1)\chi^{j_2}(\gamma_2) = \sum_{m_j=-j_1}^{j_1} \exp(-m_1\gamma_1) \sum_{m_j=-j_2}^{j_2} \exp(-m_2\gamma_2) \tag{4.39d}$$

For $\gamma_1 = \gamma_2 = \gamma$ one obtains

$$\chi^{j_1}(\gamma)\chi^{j_2}(\gamma) = \sum_{m_1 m_2} \exp[-i(m_1 + m_2)\gamma]$$

$$= \sum_{j=0}^{\infty} b(j) \sum_{m=-j}^{i} \exp(-im\gamma) \tag{4.40a}$$

where the right side is the character of $\sum b(j)\mathbf{D}^j$.

[In the following recall (4.38c).] First note that $b(j > j_1 + j_2) = 0$ as there are no such j values for given j_1 and j_2. Furthermore, $b(j_1 + j_2) = 1$ as there is only one term corresponding to $j = j_1 + j_2$ for which $m_j = j_1 + j_2$. The smallest value of j is $|j_1 - j_2|$ and for $j < |j_1 - j_2|$, $b(j) = 0$. With these properties, (4.40a) may be rewritten

$$\chi^{j_1}(\gamma)\chi^{j_2}(\gamma) = \sum_{j=|j_1-j_2|}^{j_1+j_2} \sum_{m=-j}^{j} \exp(-im\gamma) = \sum_{j=|j_1-j_2|}^{j_1+j_2} \chi^j(\gamma)$$

or, equivalently,

$$\mathbf{D}^{j_1} \otimes \mathbf{D}^{j_2} = \mathbf{D}^{j_1+j_2} + \mathbf{D}^{j_1+j_2-1} + \ldots + \mathbf{D}^{|j_1-j_2|} \tag{4.40b}$$

This sequence is noted to be parallel to the addition rule (4.38c). We note that the addition in (4.40b) is not standard as matrices on the right side are of different dimensions. It is understood that these matrices are placed along the diagonal of the direct-product matrix of dimension $(2j_1 + 1)(2j_2 + 1) \times (2j_1 + 1)(2j_2 + 1)$. With (4.40b) we note that wavefunctions of the coupled system transform as the $j = |j_1 - j_2|, \ldots, (j_1 + j_2)$ irreps of the rotation group. The dimensions of each submatrix on the diagonal is the dimension of the corresponding irrep of the rotation group and, furthermore, represents the degeneracies of corresponding eigenstates.

Consider the example of the coupling of $p(\ell = 1)$ and $d(\ell = 2)$ spinless electrons which gives $j = 1, 2, 3$. The related submatrices have dimensions $3, 5, 7$. The $\mathbf{D}^1 \otimes \mathbf{D}^2$ direct-product matrix is of dimension $(3 \times 5)^2$ so that the sub-j matrices fill the diagonal of the direct-product matrix. Corresponding eigenstates transform as the $j = 1, 2, 3$ irreps of the rotation group. Dimensions of corresponding irreps are 3, 5 and 7 which also represent degeneracies of related eigenstates.

4.5   Symmetric Group; Cayley's Theorem

Permutations

There are $n!$ permutations on a set of n objects. Each such permutation is an element of a group of order $h = n!$, called the permutation group or the symmetric group of degree n and is labeled $\mathcal{S}_n$. (This group should not be confused with S_n which denotes improper rotation; Table 1.1.) Rules of multiplication of elements of this group are as follows. Consider the group $\mathcal{S}_6$ which contains 6! elements. One such element, which we label A, is shown below,

$$A = \begin{pmatrix} 1 & 2 & 3 & 4 & 5 & 6 \\ 2 & 1 & 6 & 5 & 4 & 3 \end{pmatrix} \tag{4.41a}$$

In this permutation $1 \to 2, 2 \to 1, 3 \to 6$, etc. A shorthand notation for this permutation is as follows.

$$A = (12)(36)(45) \tag{4.41b}$$

The first of these terms corresponds to the transition $1 \to 2, 2 \to 1$, the second to $3 \to 6, 6 \to 3$ and the third to $4 \to 5, 5 \to 4$. Note that terms in each parentheses are cyclic. Thus, for example, the elements $(123), (231), (312)$ are equivalent. A permutation term in S_3 is written with respect to the ordered sequence (123). In general, permutations are described with respect to a standard sequence (of numbers or letters, etc.) In the description above, this sequence is (123456). The corresponding identity element is $E = (1)(2)(3)(4)(5)(6)$.

To describe multiplication of elements of the symmetric group consider the example of the multiplication of two elements in S_3,

$$C = \begin{pmatrix} 1 & 2 & 3 \\ 1 & 3 & 2 \end{pmatrix} \times \begin{pmatrix} 1 & 2 & 3 \\ 3 & 2 & 1 \end{pmatrix} = \begin{pmatrix} 1 & 2 & 3 \\ 2 & 3 & 1 \end{pmatrix} = (123) \tag{4.41c}$$

or, equivalently,

$$C = (1)(23) \times (13)(2) = (123) \tag{4.41d}$$

In the left product of (4.41c), multiplying from right to left, $3 \to 1, 1 \to 1; 2 \to 2, 2 \to 3; 1 \to 3, 3 \to 2$. There are $3! = 6$ elements in the group S_3, the symmetric group of degree 3 and order 6.

In this manner we find that products of elements of the permutation group are well defined and each element has an inverse. However, the following should be noted. In constructing the S_3 group not all permutations of (123) are independent. Thus, for example, as noted above, the term (123) has two other equivalent forms. The six independent elements which contribute to the S_3 group are $(1)(2)(3); (123); (321); (1)(23); (2)(31); (3)(12)$. Note that

$$(1)(23) \times (2)(31) = (312) = (123)$$

The group S_3 (of order $3! = 6$) is isomorphic to the C_{3v} group.

Here are two additional examples. The first involves elements of the S_4 group (multiplying from right to left in the middle term):

$$\begin{pmatrix} 1 & 2 & 3 & 4 \\ 4 & 1 & 3 & 2 \end{pmatrix} \times \begin{pmatrix} 1 & 2 & 3 & 4 \\ 2 & 3 & 1 & 4 \end{pmatrix} = (421)(3) \times (123)(4) = (423)(1)$$

The second involves elements of the P_6 group:

$$(153)(24)(6) \times (164)(235) = (162)(45)(3)$$

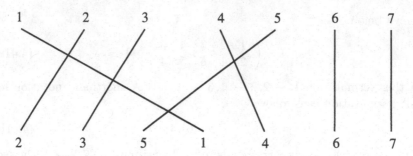

Figure 4.1. Number of 4 crossings in the diagram indicates that the permutation $(1234567) \rightarrow (2351467)$ is even.

For example, the (45) term stems from the products $5 \rightarrow 2, 2 \rightarrow 4; 4 \rightarrow 1$, $1 \rightarrow 5$.

Even and Odd Permutations; Irrep of S_n

Consider the element of S_6, (12)(345)(6). The first parentheses on the left involves one exchange of elements, the second two exchanges, and the third no exchanges. So this permutation involves three exchanges. The permutation (214536) of (123456) involves three exchanges. This is seen in the following manner. Write this permutation beneath and parallel to the original set, 1 2 3 4 5 6, and draw connecting lines joining equal numbers. You will find that the number of line crossings is 3 and the permutation is labeled an odd permutation (Fig. 4.1). These numbers of exchanges may be obtained also by counting the number of exchanges of neighboring sets of values in the primary set to obtain the given permutation. For example, the element (2 1 4 5 3 6) considered above, involves three neighboring exchanges to obtain (1 2 3 4 5 6) and is labeled an odd permutation, in accord with our previous findings.

If X_1 exchanges are involved in the permutation A_1 and X_2 exchanges in A_2, then $X_1 + X_2$ exchanges are required to describe the product permutation $A_1 A_2$. It follows that evenness or oddness of the $A_1 A_2$ permutation depends on the evenness or oddness of the sum $X_1 + X_2$. The following table applies (e for even, o for odd):

$$e \times e = e, \qquad e \times o = o,$$

$$o \times e = o, \qquad o \times o = e. \qquad (4.42)$$

One concludes that there is a one-dimensional irrep of S_n corresponding to $+1$ for the even permutations and to -1 for odd permutations. This irrep is called the antisymmetric representation. In addition, as is always the case, there is a symmetric representation composed of all 1's. Furthermore, the subgroup of S_n consisting of even permutations only, contains $n!/2$

Table 4.1

Number of Irreps, $r(n)$, of S_n for $n < 9$

n	$r(n)$
1	1
2	2
3	3
4	5
5	7
6	11
$\vdots$	$\vdots$
n	number of partitions of n

elements and is called the 'alternating group.' The alternating group is an invariant subgroup of S_n (Chapters 5 and 7).

We recall that the number of irreps of a group is equal to the number of classes of the group.

Definition. A *partition* of a positive integer is equal to the set of integers whose sum is n.

Theorem. The number of classes, $r(n)$, of S_n equals the number of partitions of n. It follows that the number of irreps of S_n equals $r(n)$. (Table 4.1. These topics are returned to below in the discussion of Young diagrams.)

Theorem. The parameter $r(n)$ is the coefficient of x^n in the power-series expansion of the

$$\text{Euler Generating function, } E(x) = \prod_{i=1}^{\infty}(1 - x^i)^{-1}$$

Cayley's Theorem

Cayley's theorem states that any finite group of order n is isomorphic to a subgroup of the symmetric group of degree n (and of order $n!$). The technique of proof of this theorem is based on the observation that by the rearrangement theorem (Section 1.3), the rows of a group table are permutations of each other. One then establishes that these permutations comprise a group which is isomorphic to the original group. This property establishes that the original group is isomorphic to a subgroup of the permutation group. The technique is demonstrated for the case that the rows of a given group table are cyclic permutations of each other (which in the present case are considered as independent elements).

Consider the cyclic subgroup of S_5 of order 5 with the group table

$$
\begin{array}{c|ccccc}
 & 0 & 1 & 2 & 3 & 4 \\
\hline
0 & 0 & 1 & 2 & 3 & 4 & E \\
1 & 1 & 2 & 3 & 4 & 0 & A \\
2 & 2 & 3 & 4 & 0 & 1 & B \\
3 & 3 & 4 & 0 & 1 & 2 & C \\
4 & 4 & 0 & 1 & 2 & 3 & D
\end{array}
\qquad (4.43a)
$$

The binary operation for this group is addition (mod 5). Permutations are identified on the left with letters shown. We wish to show that these letters form a group which has the same group table as above, thereby establishing that the two groups are isomorphic. To construct this table we note that there is a one-to-one correspondence between permutations and identifying letters and employ the permutation multiplication described above to construct group elements. Here are two examples (multiplying from right to left),

$$
AB = \begin{pmatrix} 0 & 1 & 2 & 3 & 4 \\ 1 & 2 & 3 & 4 & 0 \end{pmatrix} \begin{pmatrix} 0 & 1 & 2 & 3 & 4 \\ 2 & 3 & 4 & 0 & 1 \end{pmatrix} = \begin{pmatrix} 0 & 1 & 2 & 3 & 4 \\ 3 & 4 & 0 & 1 & 2 \end{pmatrix} = C
$$
$$(4.43b)$$

$$
AC = \begin{pmatrix} 0 & 1 & 2 & 3 & 4 \\ 1 & 2 & 3 & 4 & 0 \end{pmatrix} \times \begin{pmatrix} 0 & 1 & 2 & 3 & 4 \\ 3 & 4 & 0 & 1 & 2 \end{pmatrix} = \begin{pmatrix} 0 & 1 & 2 & 3 & 4 \\ 4 & 0 & 1 & 2 & 3 \end{pmatrix} = D
$$

Continuing in this manner, one finds

$$
\begin{array}{c|ccccc}
 & E & A & B & C & D \\
\hline
E & E & A & B & C & D \\
A & A & B & C & D & E \\
B & B & C & D & E & A \\
C & C & D & E & A & B \\
D & D & E & A & B & C
\end{array}
\qquad (4.43c)
$$

We recognize this to be the C_5 group table. It is evident that tables (4.43a) and (4.43c) exhibit identical patterns and are therefore isomorphic [as follows directly from the prime number theorem (Section 1.4)]. This demonstrates a method of proof of Cayley's theorem.

As noted above, in the group table (4.43a), number sequences in rows are cyclic replicas of each other. For consistency, multiplication of elements is given by products of column forms, as in (4.43b). The corresponding identity element is $(0)(1)(2)(3)(4)$.

Equivalent Permutations

The equivalence of cyclic permutation elements may be explicitly demonstrated with elements of the S_4 group in which cycles of the (1234) element multiply the common permutation (3214):

$$(1234) \times (3214) = A, (2341) \times (3214) = B$$

$$(3412) \times (3214) = C, (4123) \times (3214) = D$$

We find

$$A = B = C = D = (1)(2)(3)(4) = E$$

The S_3 group has six elements. Non-cyclic forms of these elements are given by $E \equiv (1)(2)(3), A \equiv (123), B \equiv (132), C \equiv (1)(23), D \equiv (2)(13), F \equiv (3)(12)$. With these labels the S_3 group table is given by

S_3	E	A	B	C	D	F
E	E	A	B	C	D	F
A	A	B	E	F	C	D
B	B	E	A	D	F	C
C	C	D	F	E	A	B
D	D	F	C	B	E	A
F	F	C	D	A	B	E

In agreement with the subgroup-divisor theorem, we see that the order, 3, of the subgroup, $C_3 = (E, A, B)$, divides the order $3! = 6$ of the S_3 group. Note also that S_3 is isomorphic to the C_{3v} group [see (1.10)]. The study of permutations is returned to in Section 7.4.

4.6 Young Diagrams[1]

Partitions

We return to the notion of the partition of a number discussed briefly in the previous section. Young diagrams provide a diagrammatic description of the partitions and provide a means of obtaining irreps of the permutation group S_n. The construction is as follows. Let n be a positive integer and let

$$\lambda_1 \geq \lambda_2 \geq \ldots \geq \lambda_h, \ \lambda_i = \text{integer} \tag{4.44a}$$

such that

$$\lambda_1 + \lambda_2 + \ldots + \lambda_h = n \tag{4.44b}$$

The sequence $[\lambda_1, \lambda_2, \ldots, \lambda_h]$ is labeled a 'partition' of n. The number 4 has the following five partitions:

$$[4], [3, 1], [2, 2] \equiv [2^2]$$

$$[2, 1, 1] \equiv [2, 1^2], \ \ [1, 1, 1, 1] \equiv [1^4] \tag{4.44c}$$

---

[1]This formalism was created by Alfred Young (1873–1940).

Numbers of Irreps of $\mathcal{S}_n$

The partitions of the integers $n = 2, 3, 4$ may be expressed as Young diagrams as follows.

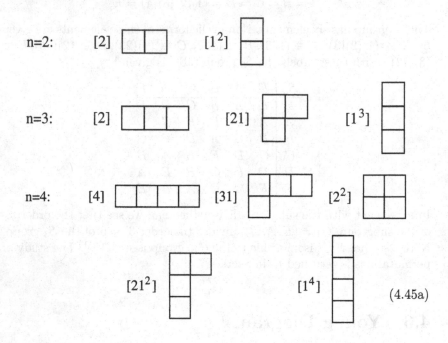

$$(4.45a)$$

Each partition, and therefore each diagram, defines a class of the $\mathcal{S}_n$ group. It follows that the number of Young diagrams for a given n value gives the number of irreps of that group. (Recall rule **V**, Section 3.4.) One may conclude that $\mathcal{S}_2, \mathcal{S}_3$ and $\mathcal{S}_4$ have two, three and five irreps, respectively. The number of boxes in a Young diagram equals the number being partitioned. If a Young diagram has a bend in it, the 'vertex' of this diagram is always at the upper left corner.

Dimensions of Irreps of $\mathcal{S}_n$

Young diagrams may also be used to determine the dimensions of irreps of a given $\mathcal{S}_n$ group. This is obtained in the following manner. For a given n, squares are numbered from 1 to n, so that numbers increase in a row from left to right and in a column from top to bottom. Such diagrams are called *standard tableaux*. The sequence of numbers (independent of their order) in any row or column appears only once in a set of congruent diagrams. The standard tableaux for $\mathcal{S}_2, \mathcal{S}_3$ and $\mathcal{S}_4$ are given by the following, together with related partition symbols:

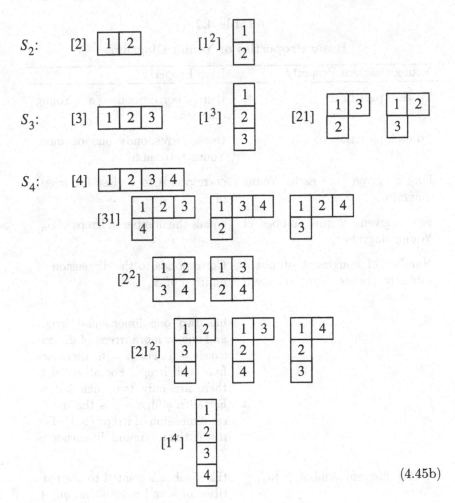

(4.45b)

Note that, for example, there are three tableaux related to the partition $[21^2]$. The dimension of an irrep of $\mathcal{S}_n$ is given by the number of congruent (with no rotation) standard tableaux related to the corresponding partition. Thus we may conclude:

$\mathcal{S}_2$ has two one-dimensional irreps corresponding to $[2]$ and $[1^2]$, respectively. (That is, there is a single standard tableaux related to $[2]$ and a single standard tableaux related to $[1^2]$.)

$\mathcal{S}_3$ has two one-dimensional irreps corresponding to $[3]$ and $[1^3]$, respectively, and one two-dimensional irrep corresponding to $[21]$.

$\mathcal{S}_4$ has two one-dimensional irreps corresponding to $[4]$ and $[1^4]$, respectively, and one two-dimensional irrep corresponding to $[2^2]$ and two three-dimensional irreps corresponding to $[31]$ and $[21^2]$, respectively. Note

Table 4.2

Basic Properties of Young Diagrams

Young Diagram Property	Irrep Property
To each partition:	there corresponds a Young diagram.
To each partition:	there corresponds one or more Young tableaux.
For a given n, each Young diagram:	corresponds to a class (or irrep) of $\mathcal{S}_n$.
For a given n, the number of Young diagrams:	equals the number of irreps of $\mathcal{S}_n$.
Number of congruent standard tableaux related to a given partition:	Corresponds to the dimension of an irrep of $\mathcal{S}_n$.
$\mathcal{S}_n$:	has two one-dimensional irreps and one or more irreps of dimension $n - 1$. For $n = 6$, there are four such irreps. For all other n there are only two such irreps. For all $n \neq 4, n - 1$ is the smallest dimension of irreps (> 1). For $n = 4$, the minimum dimension is 2.
Young diagram symbol, $[n_1^q n_2^p]$:	this symbol is related to the partition of n and is such that $qn_1 + pn_2 = n$.

that each $\mathcal{S}_n$ group has two one-dimensional irreps corresponding to the partitions $[n]$ and $[1^n]$, respectively. All other irreps of $\mathcal{S}_n$ have dimension greater than one. The preceding results are seen to be consistent with the number of irreps corresponding to $\mathcal{S}_n$ listed in Table 4.1.

To sum up: (a) The number of partitions of Young diagrams corresponding to a given value of n of the symmetric group, $\mathcal{S}_n$, is equal to the number of irreps of $\mathcal{S}_n$. (b) The dimension of an irrep related to a given partition is equal to the number of standard tableaux for the given partition (see Table 4.2).

The symmetry group is returned to in Chapter 7 in discussion of the Galois group. Young diagrams appear again in Section 6.5 in relation to irreps of the general linear group.

Young Diagrams and Wavefunctions

We discuss two applications of Young diagram representations of wave-functions. First, we recall that symmetry properties of a wavefunction for a system of identical particles stem from the Pauli principle. The diagrammatic representation further delineates the basis function for the permutation group of a given order. The Pauli principle states that: (a) The wavefunction of a collection of identical bosons (integral spin) is symmetric with respect to exchange of the spin and spatial coordinates of pairs of particles. (b) The wavefunction of a collection of identical fermions (half-odd integer spin) is antisymmetric with respect to exchange of the spin and spatial coordinates of pairs of particles. Thus, for example, the symmetric state, ψ_S, and the antisymmetric state, ψ_A, of a system of two particles are written, respectively, as

$$\psi_S = \psi(1,2) + \psi(2,1)$$

$$\psi_A = \psi(1,2) - \psi(2,1)$$

(4.45c)

In this notation '1' represents spin and space coordinates, z_1, etc., and the first slot in $\psi(,)$ is for the state of particle number 1, etc. In the state $\psi(2,1)$, particle number 1 is in the state z_2 and particle number 2 is in the state z_1. The exchange operator, X_{12}, is such that

$$X_{12}\psi(1,2) = \psi(2,1) = \pm\psi(1,2)$$

$$X_{12}\psi_S(1,2) = +\psi_S(1,2); \ X_{12}\psi_A(1,2) = -\psi_A(1,2)$$

The first equation indicates that X_{12} has two eigenvalues, ± 1, where $+1$ corresponds to an even function and -1 to an odd function.

Two-particle symmetric and antisymmetric wavefunctions may be represented as Young diagrams as follows. In this picture, each particle is associated with a box,

$$\Psi_S = \boxed{} \quad \Psi_A = \begin{array}{c}\boxed{}\\\boxed{}\end{array}$$

(4.46a)

Three-Particle Wavefunctions

For a three-particle system there are three Young diagrams:

$$\Psi_S = \begin{array}{|c|c|} \hline & \\ \hline \end{array}$$

$$\Psi_A = \begin{array}{|c|} \hline \\ \hline \\ \hline \end{array}$$

$$\Psi_M = \begin{array}{|c|c|} \hline & \\ \hline \\ \cline{1-1} \end{array}$$

$$\text{(4.46b)}$$

In this representation the third diagram (M for "mixed") represents all states which are symmetric with respect to the exchange of two particles but antisymmetric with respect to the exchange between one of the symmetrized pair and the third particle.

To find basis functions of irreps of S_n, the following operators are introduced. The *symmetrizing operator*, S, is defined as

$$S = \sum_P P \qquad (4.47a)$$

where P denotes permutation of elements in a row of a Young tableau. The *antisymmetrizing operator*, A, is given by

$$A = \sum_Q (-1)^q Q \qquad (4.47b)$$

where Q represents permutation of elements of a column of a Young tableau and q is the parity ($+$ or $-$) of Q. The product AS, relevant to a state of mixed symmetry, is labeled the 'Young operator,'

$$Y = AS \qquad (4.47c)$$

Let us apply these operators to discover the basis functions for the irreps of P_3. Consider first the row diagram of (4.44c),

$$S\psi(123) = \sum_P \psi(123) = \psi(123) + \psi(132) + \psi(213)$$
$$+\psi(231) + \psi(312) + \psi(321) = \psi_S(123) \qquad (4.47d)$$

This is the symmetric basis function for the one-dimensional symmetric irrep of S_3 and is such that $X_{ij}\psi_S = +\psi_S$, where $i \neq j$ are integers in the interval $(1,3)$. Consider next the column diagram of (4.46b). Application of the antisymmetrizing operator gives

$$A\psi(123) = \sum_Q (-1)^q Q\psi(123) = \psi(123) - \psi(132) + \psi(231)$$
$$-\psi(213) + \psi(312) - \psi(321) = \psi_A(123) \qquad (4.47e)$$

This is the antisymmetric basis function for the one-dimensional antisymmetric irrep of S_3 and is such that $X_{ij}\psi_A = -\psi_A$. There are two Young tableau corresponding to ψ_M:

$$\begin{array}{|c|c|}\hline 1 & 2 \\\hline 3 \\\cline{1-1}\end{array} \quad \begin{array}{|c|c|}\hline 1 & 3 \\\hline 2 \\\cline{1-1}\end{array}$$

$$(4.47f)$$

Consider first the tableau on the left for which the Young operation gives

$$Y\psi(123) = AS\psi(123)$$

$$S_{12}\psi(123) = \psi(123) + \psi(213) \tag{4.47g}$$

$$A_{13}S_{12}\psi(123) = \psi(123) - \psi(321) + \psi(213) - \psi(231) = \psi_3(123)$$

In like manner for the right Young tableau of (4.47f) we obtain

$$Y\psi(123) = A_{12}[\psi(123) + \psi(321)] \tag{4.47h}$$

$$= \psi(123) - \psi(231) + \psi(321) - \psi(312) = \psi_4(123) \tag{4.47i}$$

The functions ψ_3 and ψ_4 are a basis for the two-dimensional irrep of S_3.

Consider the Young tableau:

$$\begin{array}{|c|c|c|c|c|}\hline 1 & 2 & 4 & 3 & 6 \\\hline 5 \\\cline{1-1}\end{array}$$

$$(4.48a)$$

This diagram describes a wavefunction in which particles 1 and 5 are in an antisymmetric state and particles 1, 2, 3, 4, 6 are in a symmetric state. This symmetric component of the wavefunction is invariant to permutation of the five particles. However, exchanging, say, particles 6 and 5 gives a wavefunction in which particles 1 and 6 are in an antisymmetric state, and particles 1, 2, 3, 4, 5 are in a symmetric state, which is not equal to the wavefunction of the preceding tableau. This rule comes into play in the counting of Young diagrams for determining the number of irreps of S_n described above.

With reference to (4.45b) and the latter observation we note that in any Young diagram description of an n-particle system, there is always one tableau corresponding to a totally symmetric state (horizontal array) relating to a set of independent bosons, and one tableau corresponding to a totally antisymmetric state (vertical array) relating to a set of independent fermions. Each such tableau with n boxes is a one-dimensional representation of the permutation group, S_n.

Next consider the state

$$\begin{array}{|c|c|}\hline 1 & 2 \\\hline 3 & 4 \\\hline\end{array} \quad \Psi(1,2,3,4) = A_{13}\,A_{24}\,S_{12}\,S_{34}\,\Psi(1,2,3,4)$$

$$(4.48b)$$

Note that

$$
\begin{array}{|c|c|}
\hline 1 & 2 \\
\hline 3 & 4 \\
\hline
\end{array}
\;=\;
\begin{array}{|c|c|}
\hline 2 & 1 \\
\hline 4 & 3 \\
\hline
\end{array}
$$

(4.48c)

In general, the wavefunction corresponding to a square Young tableau remains invariant under exchange of rows of the tableau.

Consider the three [31] Young diagrams in (4.45b). Each of these diagrams corresponds to two particles in an antisymmetric state. The fact that there are three such independent diagrams indicates that the corresponding three functions are the basis of a three-dimensional irrep of S_4.

4.7   Degenerate Perturbation Theory

In quantum mechanical perturbation theory, the Hamiltonian is written

$$H = H_0 + H'$$

(4.49a)

where eigenfunctions and eigenenergies of H_0 are known and

$$||H'|| \ll ||H_0||$$

(4.49b)

In this relation, the norm $||x||$ maps the operator x onto a number. Approximate eigenfunctions and eigenenergies of H are sought as expansions of these eigenparameters in terms of the parameter of smallness,

$$\lambda \equiv ||H'||/||H_0||$$

(4.49c)

Degeneracies of eigenstates correspond to symmetries of H_0. If the symmetries of H_0 comprise the group G, then, as noted previously, degrees of degeneracies are equal to respective dimensions of irreps of G.

Degeneracies are removed if H' breaks the symmetries of H_0. Consider the example of a particle confined to a square-well prism centered at the origin of a Cartesian coordinate frame with the principal axis on the z axis and sides parallel to the x, y axes, respectively. The related potential function is infinite outside the prism and is zero inside the prism. The unperturbed Hamiltonian H_0 has C_{4v} symmetry and eigenstates are, in general, twofold degenerate. (As in Chapter 1, reflections in the z plane of the prism are neglected.) Figure 4.2 depicts concentrically interior perturbing potential barriers. The potentials H' in cases (a) and (b) have C_3 and C_2 symmetries, respectively, and break the symmetry of H_0. The potentials H' in cases (b) and (c) have C_4 and C_8 symmetries, respectively, and maintain the C_4 symmetry of H_0 (and hence, the degeneracy of its eigenstates). For the square quantum-well prism one has the following rule: An interior perturbing potential with C_{4n} rotational symmetry does not break the symmetry of H_0. Thus, for example, an interior hexagon perturbing potential $(6 \neq 4n)$ breaks the symmetry of H_0. Moreover, if H_0 has symmetry C_m, then any interior perturbing potential barrier with C_{km} symmetry $(k, m \geq 1$ are integers) does not break the symmetry of H_0. Equivalently, the symmetry of $H_0(C_m)$ is not broken by a prism of $C_{m'v}$ provided that

$$C_{mv} \subset C_{m'v} \tag{4.50a}$$

That is, C_{mv} is a subgroup of $C_{m'v}$, where $m' = km$, and the integer $k \geq 1$. With property (4.50a), we may say that if the group of symmetries of H_0 is C_{mv}, then a perturbing potential, concentric with and interior to the boundaries of H_0 and with symmetry $C_{m'v}$, does not break the symmetry of H_0.

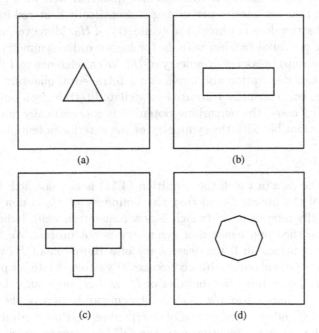

(a)   (b)

(c)   (d)

Figure 4.2. The square quantum-well prism with four perturbing interior potential barriers.

We may rewrite this symmetry-preserving criterion as follows:

$$H_0(C_{mv}) \mapsto H'(C_{m'v}) \tag{4.50b}$$

where, for the present discourse, $H_0 \mapsto H'$ indicates that the perturbing potential barrier of H' is concentrically interior to the boundary of H_0.

For three-dimensional convex polyhedra, we consider first the case that H_0 has a cubical well potential centered at the origin of a Cartesian frame with edges parallel to the x, y, z axes, respectively. With the dual properties described in Section 2.5, we may conclude that a concentrically centered interior perturbing octahedral potential-well maintains the symmetry of H_0. The same is true of any interior perturbing potential with O_h symmetry. Consider the relation

$$T_d \subset O_h \tag{4.51}$$

where, we recall, T_d is the symmetry group of the tetrahedron and O_h is the symmetry group of the cube/octahedron [see (2.16c)]. With (4.50b) and the preceding, we may conclude that a cubical or octahedral perturbation concentrically inscribed interior to a tetrahedron quantum well does not break the symmetry of $H_0(T_d)$.

An analogous description applies for the dodecahedron and its dual, the icosahedron, whose symmetry is described by the I_h group. If, say, the potential of H_0 is that of a dodecahedral quantum well with related I_h-symmetry, then an interior perturbing concentrically centered icosahedral potential barrier does not break the symmetry of H_0. Moreover, no interior perturbing potential barriers with I_h (or higher order symmetry of which I_h is a subgroup) break the symmetry of H_0. With reference to Table 2.3 we note that this description also applies to a tetrahedral quantum well with an interior, concentrically truncated or stellated tetrahedron potential. In all preceding cases, the perturbing potential is concentrically oriented in a manner compatible with the symmetry of the exterior potential.

Reciprocity

Consider the case in which the condition (4.51) is not satisfied. Reversing the roles of the potentials, so that the boundary of H_0 is now a barrier concentrically interior to H' (which is now a quantum well), indicates that (4.51) is satisfied and hence that symmetry is not broken. We then have a reciprocity principal: If the degeneracy of a Hamiltonian H_0 with rotational symmetries is broken, by a concentrically situated interior perturbing potential H', then reversing the roles of H_0 and H' maintains the symmetry of H'. A simple example of these statements is that of the spherical quantum well and any concentrically inscribed regular polyhedral potential barrier. The spherical quantum well has $O(3)$ symmetry which is broken by the interior perturbing potential. However, in the reversed case, the regular polyhedron is the potential of the unperturbed Hamiltonian and the spherical potential well becomes the interior perturbing potential barrier. In this event, the symmetry of the polyhedron quantum well is maintained.

Summary of Topics for Chapter 4

1. Degenerate eigenstates.

2. Irreps and degeneracy.

3. Representation of $O(3)$. Euler angles.

4. Homomorphism between $SU(2)$ and $O(3)^+$.

5. Irreps of $SU(2)$. Polynomial representation.

6. Irreps of $O(3)^+$.

7. Relation of $SU(2)$ and $O(3)^+$ to angular momentum.

8. Permutation group $\mathcal{S}_n$.

9. Cayley's theorem.

10. Young diagrams. Irreps of $\mathcal{S}_n$.

11. Young tableaux representations of wavefunctions.

12. Degenerate perturbation theory.

Problems

4.1 Consider two eigenfunctions, φ_1 and φ_2, which both correspond to the same eigenenergy, E, relevant to a system defined in a convex domain of volume Ω. Construct two linear combinations of these functions which are orthogonal to each other.

4.2 (a) Establish the right equality in (4.19) relevant to the characters of the irreps, D^ℓ. (b) What is the dimension of a D^ℓ element?

4.3 Establish (4.13) to discover the character of the full rotation group $O(3)$ (recall Section 2.3).

4.4 Show that if $U_1, U_2 \in SU(2)$, then $U_1 U_2 = U_3 \in SU(2)$. The matrix U is defined in (4.22).

4.5 (a) Employing (4.23) for the definition of A, A', derive the explicit form of $R(a, b)$ given in (4.24a). (b) Establish the properties (4.24b).

4.6 Show that the matrices, $U^{(j)}_{m'm}(a, b)$, (4.31b) of the irrep of $SU(2)$ are unitary.

4.7 Show that the (a, b) coefficients (4.27d) give the full rotation matrix $R(\alpha, \beta, \gamma)$.

4.8 (a) What is the product of the elements of the P_4 group

$$W = (134)(2) \times (234)(1)?$$

(b) What are the odd/even properties of the two terms in this product? What is the odd/even property of W?

4.9 Show that the group C_3 is isomorphic to a subgroup of S_3.

4.10 Employing Young diagram techniques with regard to the S_6 group: (a) Determine the number of irreps of this group. (b) Determine the dimensions of these irreps. *Hint*: For (a) first determine the partitions of 6.

4.11 A point particle is confined to a spherical cavity with perfectly reflecting walls. What is the degeneracy of a quantum state of this system corresponding to given value of ℓ?
Answer
The Hamiltonian of this system has $O(3)^+$ spherical symmetry. Irreps of this group were found to be of $2\ell + 1$ dimensions. It follows that eigenstates of given ℓ are $(2\ell + 1)$-fold degenerate.

4.12 What is the identity permutation element of S_{10}?
Answer
The identity permutation element of S_n is $(1)(2)\dots(n)$.

4.13 Write down the spin eigenstates of S^2 and S_z corresponding to a spin 1 particle.

4.14 Working in Cartesian coordinates, eigenenergies of the isotropic harmonic oscillator in 3-space vary as $E_{nqv} \propto (n+q+v)$, where (n, q, v) are positive integers ≥ 0. Writing $n+q+v = s$, the degeneracy of the (n, q, v) state is $(s+1)(s+2)/2$. If this problem is solved in spherical coordinates, eigenenergies vary as $E_{n\ell} \propto (n + 2\ell)$, where n and ℓ are positive integers ≥ 0.[2] Give an argument to show that degeneracies of the system in both frames are equal.

Answer

Degeneracies of a given system are equal to the dimensions of irreps of the group of symmetries of the system. Such irreps are independent of coordinate frames that describe the system. One may conclude that degeneracies in both coordinate frames for the isotropic harmonic oscillator are equal. In the present context, the group of symmetries that describes the three-dimensional isotropic harmonic oscillator is the $SU(3)$ group.[3] The $SU(2)$ group is a subgroup of $SU(3)$, which is consistent with the property that symmetries of the two-dimensional harmonic oscillator are described by the $SU(2)$ group. It is also noted that the sum $(n + q + v)$ is composed of all even and odd positive integers. The same is true for the sum $n + 2\ell$, so any number in the first sum is present in the second sum.

4.15 (a) State two properties of a matrix element of $SU(3)$. (b) What value of x makes the following matrix an element of $SU(3)$?

$$M = \frac{1}{2} \begin{pmatrix} 1 & i & 0 \\ -i & x & 0 \\ 0 & 0 & 1 \end{pmatrix}.$$

4.16 Consider the T_d and T_h groups. Which one of these groups is isomorphic with the symmetric group S_4? *Hint:* Consult Young diagrams.

4.17 Employing Young diagrams, show that the symmetric group, S_n, includes one irrep of dimension $n - 1$.

Answer

Consider the Young tableau given by (4.48a). Extend this diagram so that it includes numbers from 1 to n. The symmetric component of this diagram contains $n - 1$ numbers $\{x\}$. The two-particle antisymmetric component of the diagram contains the numbers (x', y),

---

[2]D. ter Haar, *Selected Problems in Quantum Mechanics*, Academic Press, New York (1964), Problem 5.5.

[3]H.L. Lipkin, *Lie Groups for Pedestrians*, North-Holland, Amsterdam, (1965), Section 4.1.

where $x' \in \{x\}$. Exchanging y with an element of $\{x\}$ ($\neq x'$) gives another tableau not equivalent to the first. Continuing this process gives $n-1$ non-equivalent tableaux, which demonstrates the existence of an irrep of S_n of dimension $n - 1$.

4.18 A point particle is confined to the interior of a dodecahedron. (a) What are the possible degeneracies of the quantum eigenstates of this system? *Hint*: Consult the character tables. (b) Describe a property of the basis functions of the one-dimensional irreps of this group.

4.19 The GOT (3.22a) indicates that matrix representations of irreps are orthogonal. Show that invariant function spaces (Λ_1, Λ_2) defined by respective basis functions of two distinct irreps (Γ_1, Γ_2) are orthogonal.

Answer

These basis functions are degenerate eigenfunctions of a Hamiltonian with symmetries which include those corresponding to the two given irreps, Γ_1 and Γ_2. Let $\varphi_1 \in \Lambda_1$ and $\varphi_2 \in \Lambda_2$ so that φ_1 and φ_2 correspond to unequal eigenvalues. In Problem 3.2 it was shown that eigenfunctions of a Hermitian operator (such as a Hamiltonian) corresponding to unequal eigenvalues are orthogonal. Thus φ_1 and φ_2 are orthogonal and we may conclude that the spaces Λ_1 and Λ_2 are orthogonal.

4.20 (a) A group G_K has four classes. Character values of the E class are $2, 1, -1, x$ and the number of elements in respective classes are $2, 2, 2, 2$. What are the values of h, x for this group? *Recall*: The number of group elements of a given class is a divisor of h. (b) Write down four matrices of this E irrep listing only diagonal elements of minimum amplitude.

4.21 Find a, b, c for which

$$(134)(25) = \begin{pmatrix} 1 & 2 & 3 & 4 & 5 \\ a & b & c & 1 & 2 \end{pmatrix}$$

4.22 Find a, c for which

$$\begin{pmatrix} 1 & 2 & 3 & 4 & 5 \\ 5 & 1 & 4 & 3 & 2 \end{pmatrix} = (1a2)(3c)$$

4.23 (a) What is the order of the S_5 group? (b) Employ Young diagrams to obtain the number of irreps of the S_5 group. (c) Employ standard tableaux to determine the dimensions of these irreps. (d) Obtain the structure of basis functions of the irreps related to the [41] partition.

4.24 (a) Derive the dimensions of irreps of the group of symmetries, $O(2)$, for the circle in the plane. Let the circle have radius a. (b) What are the basis functions for these irreps? (c) How many irreps does this group have?

Answers
(a) Repeating the derivation of these properties for the $O(3)$ group (Section 4.3), it is noted that eigenfunctions of the Laplacian, in cylindrical coordinates in the plane $z = 0$, are

$$\psi_{\ell n}(r, \theta) = J_\ell \left(\frac{x_{\ell n} r}{a} \right) \begin{pmatrix} \sin \ell\theta \\ \cos \ell\theta \end{pmatrix} \qquad \text{(P1)}$$

where

$$J_\ell(x_{\ell n}) = 0$$

and $J_\ell(kr)$ are Bessel functions of the first kind. Eigenenergies are given by $E_\ell \propto x_{\ell n}^2$. It follows that these eigenenergies are twofold degenerate. We may conclude that irreps of $O(2)$ are two-dimensional.
(b) Basis functions for an irrep corresponding to an ℓ value are given by the two solutions (P1). (c) There is an irrep for each ℓ value. It follows that there is a (countable) infinity of irreps for this group.

4.25 (a) Write down a set of elements, $\{X_i\}$, for the two-dimensional rotation group $O(2)$. (b) Show that these elements obey the four group properties (Section 1.1). (c) What is the order of this group? (d) How many irreps does this group have? (e) What are the dimensions of these irreps?
Answers (partial)
(a)

$$X_i = \begin{pmatrix} \cos \theta_i & \sin \theta_i \\ -\sin \theta_i & \cos \theta_i \end{pmatrix}$$

4.26 (a) Write down the elements of the C_4 group as a subgroup of the permutation group on four letters, S_4. (b) What is the nature of these permutations? *Hint*: For (a), label the corners of a square $(1, 2, 3, 4)$.
Answers (partial)
(b) The permutations are cyclic.

4.27 Show directly that the number of irreps of S_5 is 7.
Answer
There are 7 partitions of 5 namely,
$(5), (4, 1), (3, 2), (3, 1, 1), (2, 2, 1), (2, 1, 1, 1) (1, 1, 1, 1, 1)$

4.28 (a) Write down the Young diagrams corresponding to the S_6 permutation group.
(b) How many irreps does this group have?
(c) What are the respective dimensions of these reps?
Answers (partial)
(a) The Young diagrams corresponding to the S_6 group are given by the following:

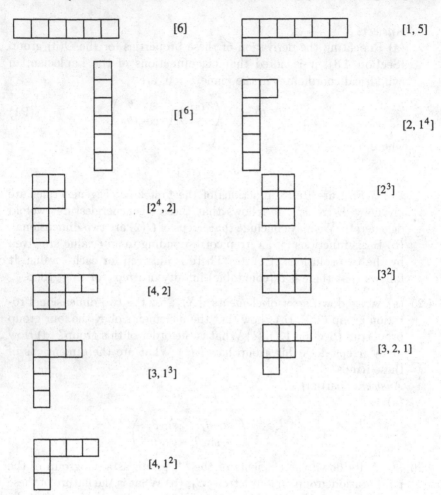

It follows that the $\mathcal{S}_6$ group of order 720 has 11 irreps.

5

Space Groups, Brillouin Zone and the Group of k

5.1 Cosets and Invariant Subgroups. The Factor Group

Consider that a finite group G has the subgroup S. We recall that the order of S divides the order of G, so that $g/s = n$, where n is an integer and g and s are the orders of G and S, respectively (Section 1.4). Let p be an element of G but not of S, so that $p \in G - S$ (Fig. 5.1).

A *complex* is a set of elements of a group. It is not necessarily a group. The complexes pS and Sp are called the *left and right cosets* of S, respectively. As the number of elements in $G - S$ is finite and unique, for a given subgroup S, the coset description is unique, The coset pS (or Sp) is not a subgroup. This follows since $E \in S$, so that $E \notin pS$. The inverse of $k \in S$ is contained in S. Two left (or right) cosets of S are either identical or disjoint (Problem 5.1).

Invariant Subgroup

The subgroup S is an invariant (or *normal* or *self-conjugate*) subgroup of G if

$$G_k^{-1} S G_k = S \tag{5.1}$$

for all $G_k \in G$. In other words, an invariant subgroup of G contains whole classes of G. Thus, for example, $\{E, C_3, C_3^2\}$, whose elements are whole classes, is an invariant subgroup of C_{3v} (1.9) whereas $\{E, 3C_2\}$ is an invariant subgroup of the cubic O group (2.13b). The right and left cosets of an invariant subgroup are equal. This follows from (5.1) as $SG_k = G_kS$.

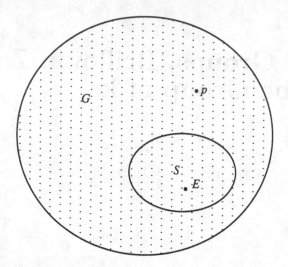

Figure 5.1. Venn diagram for the group G, subgroup S, the element $p \in G - S$ and the identity $E \in S \subset G$. If every element of S is a whole class, then S is an invariant subgroup.

The group G may be written as a *factored set* in terms of the union of its cosets with respect to the subgroup S,

$$G = S \cup pS \cup qS \ldots \tag{5.2}$$

There are $s(n-1) = g - s$ elements in the set $(G - S)$. It follows that the number of elements on the right of (5.2) is $s(n-1) + s = ns = g$.

Factor Group[1]

A *factor group* (or *quotient group*) is composed of an invariant subgroup and all its cosets. Thus, if S is an invariant subgroup, then the factor group of G with respect to the invariant subgroup S is given by the following set of complexes, labeled G/S,

$$G/S = \{S, pS, qS, \ldots\} \tag{5.3}$$

Each complex in this set is a single element of the factor group. Let us demonstrate the group properties of G/S. As S is an invariant subgroup, the right and left cosets of S are equal, so we will consider the elements pS of G/S,

$$\text{Closure: } (rS)(qS) = rSqS = rSSq = rSq = vS \tag{5.4a}$$

---

[1]The factor group was introduced by E´variste Galois (1811–1832). See Chapter 7.

$$\text{Identity:} \quad S(qS) = qSS = qS \tag{5.4b}$$

$$\text{Inverse:} \quad (qS)^{-1}qS = S^{-1}q^{-1}qS = S^{-1}S = E \tag{5.4c}$$

(In the latter equality of (5.4a), $rq = v \in G - S$ or $v \in S$. As S is a group, $S^2 = S$. In either case, closure is satisfied.)

$$\text{Associativity:} \; (pS)(qStS) = (pS)(qSt) = pSqSt = pqSt$$
$$\tag{5.4d}$$
$$(pSqS)(tS) = (pSq)(tS) = pSqtS = pqSt$$

The relations (5.4b,c) indicate, respectively, that:

(a) The invariant subgroup, S, of G/S is the identity element of this group.

(b) Each element of G/S is its own inverse.

Examples

As an example of a factor group consider the group C_{2v} (1.9) with the invariant subgroup, $S = \{E, C_2\}$. With reference to the group table (1.9), cosets of S are given by

$$\sigma_v^x S = \{\sigma_v^x E, \sigma_v^x C_2\} = \{\sigma_v^x, \sigma_v^y\}$$

$$\sigma_v^y S = \{\sigma_v^y E, \sigma_v^y C_2\} = \{\sigma_v^y, \sigma_v^x\} \tag{5.5a}$$

It follows that the factor group may be written

$$C_{2v}/S = [\{E, C_2\}, \{\sigma_v^x, \sigma_v^y\}] \equiv \{S, Q\} \tag{5.5b}$$

The group table of this set is given by (1.5a), so that C_{2v}/S is isomorphic to the C_2 group. The group C_2 has two irreps, $\Gamma_1 = 1, 1; \Gamma_2 = 1, -1$. These irreps apply as well to the full C_{3v} group, for which $\Gamma_1 = 1, 1, 1, 1; \Gamma_2 = 1, 1, -1, -1$. The product, SQ, for example, gives $\{E, C_2\}\{\sigma_v^x, \sigma_v^y\} = \{\sigma_v^y \sigma_v^x\} = Q$.

A second example addresses the C_{3v} group (1.10) with the invariant subgroup, $S = \{E, C_3, C_3^2\}$, with cosets $\{S\sigma_a, S\sigma_b, S\sigma_c\}$. Consider the products

$$(E, C_3, C_3^2)\sigma_a = (\sigma_a, \sigma_b, \sigma_c), \text{etc.} \tag{5.6a}$$

It follows that the factor group for the C_{3v} group is given by

$$C_{3v}/S = [\{E, C_3, C_3^2\}, \{\sigma_a, \sigma_b, \sigma_c\}] \equiv [S, Q] \tag{5.6b}$$

which again is isomorphic to the C_2 group. Applying irreps of this group to the C_{3v} group indicates the one-dimensional irreps: $(1, 1, 1, 1, 1, 1)$; $(1, 1, 1, -1, -1, -1)$, see (3.18a). In general, we note the rule

$$\text{irreps of } G/S \text{ correspond to irreps of } G \tag{5.6c}$$

Generalizing the preceding results we note that the group C_{nv} has a factor group

$$C_{nv}/S = [\{E, C_n, C_n^2, \ldots, C_n^{n-1}\}, \{\sigma_1, \sigma_2, \sigma_3, \ldots, \sigma_n\}] \equiv [S, Q]$$

where σ_i is a reflection element of C_{nv}. It follows that all C_{nv} groups have a two-element factor group which is isomorphic to C_2 with the group table

C_2	S	Q
S	S	Q
Q	Q	S

This table tells us that if $s \in S$ and $q \in Q$ then $sq \in Q$, $qq' \in S$ and $ss' \in S$. [See also Problem 5.6.]

Order of a Factor Group

The order of a factor group follows from (5.2),

$$h(G) = h(S)h(G/S) \tag{5.7a}$$

$$h(G/S) = \frac{h(G)}{h(S)} = \frac{g}{s} \tag{5.7b}$$

It follows that $h(G/S)$ may be far smaller than the order, g, of the full group G. In the first example cited above, $h(C_{2v}) = 4$, and $h(S) = 2$, so that, with (5.7b), $h(C_{2v}/S) = 2$. In the second example, $h(C_{3v}) = 6, h(S) = 3$, and again, $h(C_{3v}/S) = 2$. As there is only group of order 2, both C_{2v}/S and C_{3v}/S are isomorphic to C_2.

These concepts of cosets, invariant subgroups and factor groups are returned to in Section 7.1 in discussion of the ideal of a ring.

5.2   Primitive Vectors. Braviais Lattice. Reciprocal Lattice Space

Translation Vector

In the study of crystalline solids one is concerned with the concept of a Bravais lattice. A Bravais lattice is a regular array of points in space with the property that the environment about each lattice point is identical. It is defined by three *primitive translation vectors*, a_1, a_2, a_3, such that any two points in the lattice are related by

$$r' = r + a_1 n_1 + a_2 n_2 + a_3 n_3 \tag{5.8a}$$

where

$$\mathbf{T_n} = a_1 n_1 + a_2 n_2 + a_3 n_3 \tag{5.8b}$$

is the *translation vector* and n_1, n_2 and n_3 are integers. An example of these relations in two dimensions is shown in Fig. 5.2.

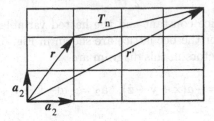

Figure 5.2. The translation vector in two dimensions.

Every point in the lattice may be reached by application of $\mathbf{T_n}$. Three fundamental Bravais lattices are the (a) simple cubic (sc) lattice, (b) body-centered cubic (bcc) lattice, (c) face-centered cubic (fcc) lattice. Note that including points at the mid-points of the edges of an sc lattice give a periodic array, but not a Bravais lattice.

Primitive Cell

A primitive cell has the three properties:

(a) It is of minimum volume.

(b) With the translation vector, the primitive cell generates the lattice.

(c) There is one lattice point per primitive cell.

The primitive cell of a given lattice is not unique. In the two-dimensional example shown in Fig. 5.3 both cells generate the lattice and both cells have the same 'volume' and each contains one lattice point.

In three dimensions the volume of a primitive cell is given by

$$V = \mathbf{a}_1 \cdot (\mathbf{a}_2 \times \mathbf{a}_3) \tag{5.9}$$

Primitive vectors for the fcc lattice are shown in Fig. 5.4.

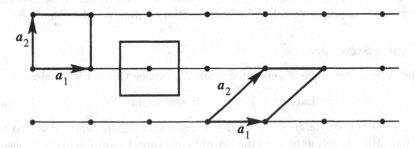

Figure 5.3. Two primitive cells for a two-dimensional array.

The primitive vectors in this diagram are

$$\mathbf{a}_1 = \frac{1}{2}a(\hat{\mathbf{x}} + \hat{\mathbf{y}}), \quad \mathbf{a}_2 = \frac{1}{2}a(\hat{\mathbf{y}} + \hat{\mathbf{z}}), \quad \mathbf{a}_3 = \frac{1}{2}a(\hat{\mathbf{x}} + \hat{\mathbf{z}}) \tag{5.10a}$$

where a is the lattice constant and the hatted variables are unit vectors. Primitive vectors for the bcc lattice are shown in Fig. 5.5.

The primitive vectors in this diagram are

$$\mathbf{a}_1 = \frac{1}{2}a(\hat{\mathbf{x}} + \hat{\mathbf{y}} - \hat{\mathbf{z}}), \quad \mathbf{a}_2 = \frac{1}{2}a(-\hat{\mathbf{x}} + \hat{\mathbf{y}} + \hat{\mathbf{z}})$$

$$\mathbf{a}_3 = \frac{1}{2}a(\hat{\mathbf{x}} - \hat{\mathbf{y}} + \hat{\mathbf{z}}) \tag{5.10b}$$

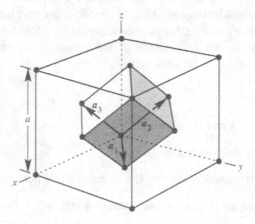

Figure 5.4. Primitive vectors for the fcc lattice and related primitive cell. (From: Kittel, *Introduction to Solid State Physics*, 7th ed., Copyright ©1995 John Wiley & Sons, Inc. Reprinted by permission of John Wiley & Sons, Inc.)

The *Wigner-Seitz* primitive cell has one lattice point per cell. It is constructed as follows. From any point in the lattice, lines are drawn to the nearest neighbors of the point. Normally bisecting planes to these lines are inserted. The space enclosed by the intersection of these planes defines the Wigner-Seitz primitive cell. This construction is illustrated in Fig. 5.6.

Crystal Structure

The crystal structure is generated from the lattice through the relation

$$\text{Lattice} + \text{Basis} = \text{Crystal Structure} \tag{5.11a}$$

The *basis* is a single group of atoms or molecules which, when attached to every lattice point, generates the crystal structure. Examples of some bases and corresponding crystal structures are shown below:

Lattice	Material (Basis)	Basis Coordinates
fcc	NaCl, LiH, MgO, AgBr, KCl	$\{(0,0,0), (1/2, 1/2, 1/2)\}$
sc	CsCl, NH$_4$Cl, CuZn (β-brass)	$\{(0,0,0), (1/2, 1/2, 1/2)\}$

For example, in single crystal NaCl, a Cl$^-$ ion is at $(0,0,0)$ and a Na$^+$ ion is at $(1/2, 1/2, 1/2)$.

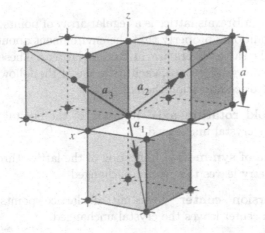

Figure 5.5. Primitive vectors for the bcc lattice and the lattice constant a. (From: Kittel, *Introduction to Solid State Physics*, 7th ed., Copyright ©1995 John Wiley & Sons, Inc. Reprinted by permission of John Wiley & Sons, Inc.)

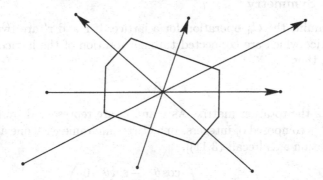

Figure 5.6. The Wigner-Seitz primitive cell for a two-dimensional lattice.

As there is one lattice point per primitive cell and there is one basis per lattice point, there is one basis per primitive cell. If the volume per primitive cell is v_p, then the number of primitive cells per volume is $n_p = 1/v_p$. The total number of bases, N_b, or the total number of primitive cells, N_p, in the sample volume V, are then given by

$$n_p V = N_b = N_p \qquad (5.11b)$$

5.3 Crystallographic Point Groups and Reciprocal Lattice Space

Bravais Lattice

As noted above, a Bravais lattice is a regular array of points that fills space and has the additional property that the environment about each point of the lattice is the same. There are 14 Bravais lattices. These may be separated into seven crystal groups, each with one of the following symmetric characteristics, or combination thereof.

(a) $[C_n]$ n-**fold rotation axis**. Rotation of the crystal through $2\pi/n$ leaves the crystal unchanged.

(b) $[\sigma]$ **Plane of symmetry**. Reflection of the lattice through the plane of symmetry leaves the crystal unchanged.

(c) $[i]$ **Inversion center**. Inverting all lattice points through the inversion center leaves the crystal unchanged.

(d) $[iC_n]$ **Rotation-inversion axis**. Rotation about an axis through $2\pi/n$ followed by inversion through a point on the axis leaves the crystal unchanged.

Rotation Symmetry

Let us examine the C_n operation for a lattice. If $\mathbf{r}$ and $\mathbf{r}'$ are two vectors of the lattice which are connected through rotation of the lattice about a fixed axis, then

$$\mathbf{r} = \mathbf{R}\mathbf{r}' \qquad (5.12a)$$

where $\mathbf{R}$ is the rotation matrix. As $\mathbf{r}$ and $\mathbf{r}'$ are represented by primitive vectors, $\mathbf{R}$ is composed of integers. In a Cartesian frame with one axis taken as the rotation axis [recall (3.11)],

$$\mathbf{R} \rightarrow \mathbf{R}' = \begin{pmatrix} \cos\theta & -\sin\theta & 0 \\ \sin\theta & \cos\theta & 0 \\ 0 & 0 & 1 \end{pmatrix} \qquad (5.12b)$$

Noting that $\mathbf{r}$ and $\mathbf{r}'$ are related through a linear transformation, if follows that

$$Tr\mathbf{R} = Tr\mathbf{R}' = \text{integer}$$

which, incorporated with the relation $-1 \leq \cos\theta \leq 1$, gives

$$Tr\,\mathbf{R}' = 1 + 2\cos\theta = -1, 0, 1, 2, 3 \qquad (5.12c)$$

Equivalently, one may write

$$\theta = 2\pi, 2\pi/2, 2\pi/3, 2\pi/4, 2\pi/6 \qquad (5.12d)$$

corresponding, respectively, to the identity, twofold, threefold, fourfold rotation and sixfold rotations. There are no other invariant rotational symmetries of a lattice. Thus, only these values of the rotation index, n, enter in the symmetry category (a) above.

Notation

In describing lattices in three-dimensional space, the following conventions are noted. The conventional cell of a Bravais lattice is a parallelepiped with six faces and eight vertices (or corners). Thus, three angles and three lengths specify each lattice. The three angles of a vertex are labeled α, β, γ and the three lengths stemming from a vertex are labeled a, b, c. If the conventional cell has only its eight vertex lattice points, the lattice is labeled P (for primitive). The symbol R denotes rhombohedral primitive. If the cell has a lattice point at its center it is labeled I (for body centered). If the cell has lattice points at the centers of its six faces, it is labeled F (for face centered). If the cell has lattice points at the centers of its base and top face it is labeled C (for base centered). Here is a list of symbols and their meanings for the Bravais lattices.

Symbols for the Bravais Lattices

Symbol	Interpretation
α, β, γ	Angles
a, b, c	Lengths
P (primitive)	Conventional cell has only its eight vertex lattice points
I (body centered)	Conventional cell includes a lattice point at its centers
F (face centered)	Conventional cell includes lattice points at its face centers
R	Rhombohedral primitive

With this notation the 14 Bravais lattices are listed in Table 5.1 and shown in Table 5.2. Note that you should be able to extend the figure for the hexagonal P lattice to a right-hexagonal cylinder.

Holohedral Groups

The primitive cell is defined by three lengths and three angles. The most general cell is the triclinic one with all values of these parameters different. The symmetries of the triclinic cell are described by the C_i group. As the angles and lengths of the cell are varied, with two or more of these parameters set equal, eventually letting all angles and lengths become equal, seven symmetry types are generalized. These are called the holohedral groups, i.e., the highest point groups compatible with the Bravais lattice.

Table 5.1

Seven Crystal Systems for the Bravais Lattices

System	Cell Angles and Dimensions	Holohedral Point Groups	Bravais Lattice
Triclinic	$a \neq b \neq c$ $\alpha \neq \beta \neq \gamma$	C_i	P
Monoclinic	$a \neq b \neq c$ $\alpha = \gamma = \pi/2 \neq \beta$	C_{2h}	P, I
Orthorombic	$a \neq b \neq c$ $\alpha = \beta = \gamma = \pi/2$	D_{2h}	P, C, I, F
Tetragonal	$a = b \neq c$ $\alpha = \beta = \gamma = \pi/2$	D_{4h}	P, I
Cubic	$a = b = c$ $\alpha = \beta = \gamma = \pi/2$	O_h	P, I, F
Trigonal	$a = b = c$ $2\pi/3 > \alpha = \beta = \gamma \neq \pi/2$	D_{3h}	R
Hexagonal	$a = b \neq c$ $\alpha = \beta = \pi/2, \gamma = 2\pi/3$	D_{6h}	P

Reciprocal Lattice Space

Let $n(\mathbf{r})$ denote electron number density in a given crystal. This parameter is periodic with periods given by the primitive vectors $\mathbf{a}_1, \mathbf{a}_2, \mathbf{a}_3$, so that

$$n(\mathbf{r} + \mathbf{T}) = n(\mathbf{r}) \qquad (5.13)$$

where $\mathbf{T}$ is the translation vector (5.8b). This periodicity permits the following Fourier representation of the electron number density.

Table 5.2
The 14 Bravais Lattices

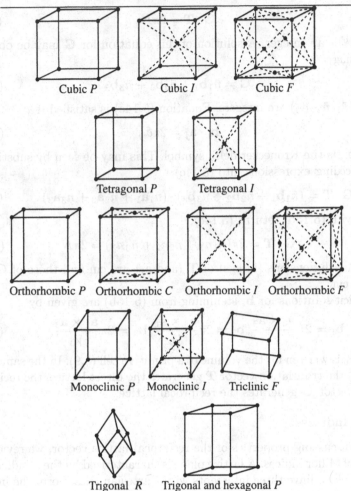

Cubic *P*   Cubic *I*   Cubic *F*

Tetragonal *P*   Tetragonal *I*

Orthorhombic *P*   Orthorhombic *C*   Orthorhombic *I*   Orthorhombic *F*

Monoclinic *P*   Monoclinic *I*   Triclinic *F*

Trigonal *R*   Trigonal and hexagonal *P*

(From: Kittel, *Introduction to Solid State Physics*, 3rd ed., Copyright ©1968 John Wiley & Sons, Inc. Reprinted by permission of John Wiley & Sons, Inc.)

$$n(\mathbf{r}) = \Sigma_{\mathbf{G}} n_{\mathbf{G}} \exp(i\mathbf{G} \cdot \mathbf{r})$$

$$= n(\mathbf{r} + \mathbf{T}) = \Sigma_{\mathbf{G}} n_{\mathbf{G}} \exp i\mathbf{G} \cdot (\mathbf{r} + \mathbf{T}) \qquad (5.14a)$$

$$= \Sigma_{\mathbf{G}} n_{\mathbf{G}} \exp i(\mathbf{G} \cdot \mathbf{r}) \exp i(\mathbf{G} \cdot \mathbf{T})$$

where $n_{\mathbf{G}}$ is the Fourier expansion coefficient for $n(\mathbf{r})$. It follows that

$$\exp i\mathbf{G} \cdot \mathbf{T} = 1 \qquad (5.14b)$$

This relation serves to determine the Fourier expansion vector, $\mathbf{G}$, namely, the preceding implies

$$\mathbf{G} \cdot \mathbf{T} = 2\pi N \qquad (5.14c)$$

where N is an integer. A solution of this equation for $\mathbf{G}$ may be obtained by writing

$$\mathbf{G} = \bar{n}_1 \mathbf{b}_1 + \bar{n}_2 \mathbf{b}_2 + \bar{n}_3 \mathbf{b}_3 \qquad (5.15a)$$

where $(\bar{n}_1, \bar{n}_2, \bar{n}_3)$ are integers. Equation (5.14c) is satisfied if

$$\mathbf{b}_i \cdot \mathbf{a}_j = 2\pi \delta_{ij} \qquad (5.15b)$$

where δ_{ij} is the Kronecker delta symbol. This may be seen by substituting the preceding expressions into (5.14c),

$$\mathbf{G} \cdot \mathbf{T} = (\bar{n}_1 \mathbf{b}_1 + \bar{n}_2 \mathbf{b}_2 + \bar{n}_3 \mathbf{b}_3) \cdot (n_1 \mathbf{a}_1 + n_2 \mathbf{a}_2 + n_3 \mathbf{a}_3) \qquad (5.15c)$$

which with (5.15b) returns (5.14c),

$$\mathbf{G} \cdot \mathbf{T} = 2\pi (\bar{n}_1 n_1 + \bar{n}_2 n_2 + \bar{n}_3 n_3) = 2\pi N \qquad (5.15d)$$

Note that dimensions (in cgs) of $\mathbf{T}$ (and $\mathbf{a}_i$) are cm and those of $\mathbf{G}$ (and $\mathbf{b}_i$) are cm^{-1}.

Explicit solutions for $\mathbf{b}_i$ stemming from (5.15b) are given by

$$\mathbf{b}_1 = 2\pi \frac{\mathbf{a}_2 \times \mathbf{a}_3}{V_0}, \, \mathbf{b}_2 = 2\pi \frac{\mathbf{a}_3 \times \mathbf{a}_1}{V_0}, \, \mathbf{b}_3 = 2\pi \frac{\mathbf{a}_1 \times \mathbf{a}_2}{V_0} \qquad (5.15e)$$

where V_0 is written for the volume of a primitive cell (5.9). In the same manner that the translation vector $\mathbf{T}$ generates the direct lattice, the reciprocal lattice vector $\mathbf{G}$ generates the reciprocal lattice.

Miller Indices

Before discussing properties of the reciprocal lattice vector, we review the notion of Miller indices. A lattice plane is characterized by these indices. Let a given plane have three intercepts with Cartesian axes. Form the inverses of these displacements and reduce these to a factored set of the smallest integers having the same ratios. Integers of the factored set gives the three Miller indices. For example, the intercepts $(3, 2, 2)$ give the reciprocals $(1/3, 1/2, 1/2) = 1/6(2, 3, 3)$. It follows that the Miller indices for the given plane are $(2, 3, 3)$, conventionally written (233). Miller indices carry the notation (hkl). A vector with components hkl is denoted as $[hkl]$.

Two Properties of G

Two principal properties of the reciprocal lattice vector ('rlv') are as follows:

(a) Every rlv is normal to a lattice plane (hkl). Thus one may write

$$\mathbf{G}(hkl) = h\mathbf{b}_1 + k\mathbf{b}_2 + l\mathbf{b}_3 \qquad (5.16a)$$

(b) Spacing between consecutive (hkl) lattice planes is given by

$$d(hkl) = \frac{1}{|\mathbf{G}(hkl)|} \qquad (5.16b)$$

For a cubic crystal with lattice constant a, one obtains

$$d(hkl) = \frac{a}{\sqrt{h^2 + k^2 + l^2}} \qquad (5.16c)$$

An rlv exists in reciprocal lattice space. However, with (5.15e) we note that $\mathbf{b}_i$ reciprocal lattice vectors are defined with respect to primitive $\mathbf{a}_i$ vectors in direct space. Thus it is consistent to consider $\mathbf{G}$ a vector in direct space as well.

Scattering Amplitude and Bragg Scattering Conditions

As the lattice constant of a typical lattice is the order of Å (10^{-8} cm), X rays incident on the sample will give interference effects. Let plane waves of X rays with propagation vector $\mathbf{k}$ be incident on a crystal. A measure of the radiation scattered in the $\mathbf{k}'$ direction is given by the *scattering amplitude*

$$F(\Delta\mathbf{k}) - \int_{xtal} d\mathbf{r}\, n(\mathbf{r}) \exp i\mathbf{r} \cdot (\mathbf{k} - \mathbf{k}') \qquad (5.17a)$$

$$\Delta\mathbf{k} \equiv \mathbf{k}' - \mathbf{k} \qquad (5.17b)$$

The relation (5.17a) reflects the property that scattering is dominant from domains of high electron density. Substituting the Fourier representation (5.14a) of $n(\mathbf{r})$ into the preceding gives

$$F(\Delta\mathbf{k}) = \Sigma_{\mathbf{G}} \int_{xtal} d\mathbf{r}\, n_{\mathbf{G}} \exp i\mathbf{r} \cdot (\mathbf{G} - \Delta\mathbf{k}) \qquad (5.17c)$$

It follows that when $\mathbf{G}$ is equal to the value $\Delta\mathbf{k}$, $F(\Delta\mathbf{k})$ is maximum. That is, when

$$\Delta\mathbf{k} = \mathbf{G} \qquad (5.18a)$$

$$F(\Delta\mathbf{k}) \to F(G) = \int_{xtal} d\mathbf{r}\, n_{\mathbf{G}} = V n_{\mathbf{G}} \qquad (5.18b)$$

which is a maximum of $F(\Delta\mathbf{k})$. The values of $\Delta\mathbf{k}$ given by (5.18a) correspond to maxima in the interference pattern of the scattered X rays.

Bragg Scattering Equations

With (5.17b) we write

$$\mathbf{k}' = \mathbf{k} + \Delta\mathbf{k} \qquad (5.19a)$$

For elastic scattering

$$k'^2 = k^2 \tag{5.19b}$$

Inserting the condition for constructive scattering (5.18a) into (5.19a) gives

$$\mathbf{k}' = \mathbf{k} + \Delta\mathbf{k} = \mathbf{k} + \mathbf{G} \tag{5.19c}$$

$$(\mathbf{k} + \mathbf{G})^2 = \mathbf{k}'^2 = \mathbf{k}^2 \tag{5.19d}$$

Completing the square gives the two Bragg scattering conditions

$$G^2 + 2\mathbf{k} \cdot \mathbf{G} = 0 \tag{5.20a}$$

$$2\mathbf{k} \cdot \mathbf{G} = G^2 \tag{5.20b}$$

In the latter relation it is noted that if $\mathbf{G}$ is an rlv so is $-\mathbf{G}$. These relations may be used to obtain Bragg's interference equation familiar from elementary physics (Problem 5.10)

$$2d\sin\theta = n\lambda \tag{5.20c}$$

where θ is the angle between a scattering plane and the incident $\mathbf{k}$ vector, d is the distance between scattering planes and n is an integer.

Brillouin Zone

We return to Bragg's second equation (5.20b) and write

$$\left(\frac{G}{2}\right)^2 = \mathbf{k} \cdot \frac{\mathbf{G}}{2} \tag{5.21a}$$

This equation has geometrical meaning (Fig. 5.7).

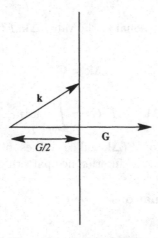

Figure 5.7. The vertical plane bisects a $\mathbf{G}$ vector.

From the figure it is evident that every **k** vector which terminates on the bisecting plane of **G** satisfies the Bragg condition (5.21a) for constructive interference.

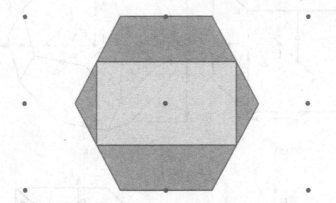

Figure 5.8. First and second Brillouin zones for two-dimensional rectangular reciprocal lattice space. First Brillouin zone is lightly shaded and the second Brillouin zone is more darkly shaded.

We are now prepared to construct the first Brillouin zone. This construction is based on the property that points in reciprocal lattice space are separated by **G** vectors. Pick a reciprocal lattice point. Draw lines to nearest neighbor reciprocal lattice points. The smallest enclosure formed by the bisecting planes of these lines is the first Brillouin zone. These bisecting planes are sometimes called 'Bragg planes.' Recalling the construction of the Wigner-Seitz cell (Fig. 5.6) we see that this cell represents the first Brillouin zone (in two dimensions). The second Brillouin zone is formed by the smallest enclosure of bisecting planes to the second nearest neighbor reciprocal lattice points. This method is shown in Fig. 5.8 for a two-dimensional lattice.

In three dimensions we note the following rules. The first Brillouin zone for an fcc lattice is the Wigner-Seitz cell in a bcc reciprocal lattice. This gives a truncated octahedron (Fig. 5.9(a)). The first Brillouin zone for a bcc lattice is the Wigner-Seitz cell in an fcc reciprocal lattice (Fig. 5.9(b)). Each successive zone has the same volume as the first.

5.4   Bloch Waves and Space Groups

Bloch Waves

Consider the time-independent Schrödinger equation for an electron propagating in a lattice. The force on the electron is represented by the periodic

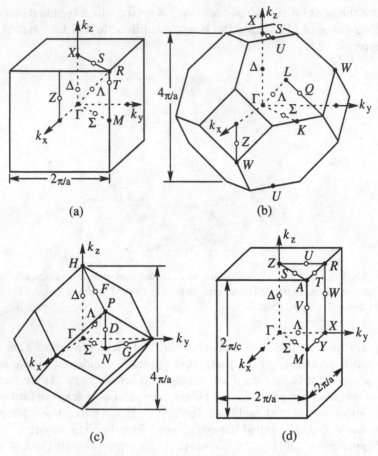

Figure 5.9. Brillouin zones for: (a) simple cubic, (b) face-centered cubic, (c) body-centered cubic and (d) simple tetragonal lattices. Also shown are conventional labels for locations and directions. (From: T. Inui, Y. Tanabe, Y. Onodera, *Group Theory and its Application in Physics*, Springer–Verlag (1966).)

potential $V(\mathbf{r}) = V(\mathbf{r} + \mathbf{T})$, where $\mathbf{T}$ is a translation vector of the lattice,

$$\left[-\frac{\hbar^2 \nabla^2}{2m} + V(\mathbf{r}) - E \right] \varphi(\mathbf{r}) = 0 \tag{5.22a}$$

As $V(\mathbf{r})$ is periodic, for the modulus of $\varphi(\mathbf{r})$ we write

$$|\varphi(\mathbf{r})| = |\varphi(\mathbf{r} + \mathbf{T})| \tag{5.22b}$$

We examine solutions of the form

$$\varphi_{\mathbf{k}}(\mathbf{r}) = u_{\mathbf{k}}(\mathbf{r}) e^{i \mathbf{k} \cdot \mathbf{r}} \tag{5.22c}$$

These forms are called *Bloch waves*. Substituting (5.22c) into (5.22a) gives

$$\left[-\frac{\hbar^2\nabla^2}{2m} + \frac{\hbar}{m}\mathbf{k}\cdot\mathbf{p} + V\right]u_{\mathbf{k}}(\mathbf{r}) = \left[E(\mathbf{k}) - \frac{\hbar^2 k^2}{2m}\right]u_{\mathbf{k}}(\mathbf{r}) \qquad (5.23a)$$

$$\mathbf{p} = -i\hbar\nabla \qquad (5.23c)$$

Again from the periodicity of $V(\mathbf{r})$,

$$u_{\mathbf{k}}(\mathbf{r}) = u_{\mathbf{k}}(\mathbf{r} + \mathbf{T}) \qquad (5.23c)$$

A consistency requirement on this construction is that in the limit that V is constant, $\varphi_{\mathbf{k}}(\mathbf{r})$ must reduce to free-article plane waves. Setting V and $u_{\mathbf{k}}$ equal to respective constants in (5.23a) gives

$$\left[E - \frac{\hbar^2 k^2}{2m} - V\right]u_{\mathbf{k}}(\mathbf{r}) = 0 \qquad (5.24a)$$

which, with $u_{\mathbf{k}} = A \neq 0$, gives

$$E = V + \frac{\hbar^2 k^2}{2m} \qquad (5.24b)$$

and the plane wave

$$\varphi_{\mathbf{k}}(\mathbf{r}) = Ae^{i\mathbf{k}\cdot\mathbf{r}} \qquad (5.24c)$$

The latter two findings are relevant to the case of an electron propagating in a constant potential field.

As $e^{i\mathbf{G}\cdot\mathbf{T}} = 1$ [see (5.14b)], we note that

$$e^{i\mathbf{k}\cdot\mathbf{T}} = e^{i(\mathbf{k}+\mathbf{G})\cdot\mathbf{T}} \qquad (5.25)$$

The eigenenergy $E(\mathbf{k})$ is likewise invariant to translations in reciprocal lattice space and we write

$$E(\mathbf{k}) = E(\mathbf{k} + \mathbf{G}) \qquad (5.26)$$

Let us ascertain the translational property of $\varphi_{\mathbf{k}}(\mathbf{r})$. With (5.22b,c) we write

$$\varphi_{\mathbf{k}}(\mathbf{r} + \mathbf{T}) = e^{i\mathbf{k}\cdot\mathbf{T}}e^{i\mathbf{k}\cdot\mathbf{r}}u_{\mathbf{k}}(\mathbf{r} + \mathbf{T}) = e^{i\mathbf{k}\cdot\mathbf{T}}\varphi_{\mathbf{k}}(\mathbf{r}) \qquad (5.27a)$$

so that the effect of translation through a translation vector is to modify the wavefunction by the phase factor $e^{i\mathbf{k}\cdot\mathbf{T}}$. With (5.27a) we write

$$|\varphi_{\mathbf{k}}(\mathbf{r} + \mathbf{T})| = |\varphi_{\mathbf{k}}(\mathbf{r})| \qquad (5.27b)$$

Electron probability density is periodic with respect to the translation vector.

Seitz Operator

Consider the combined operation of a point-group operator, R (proper or improper rotation or reflection) and a translation, $\mathbf{t}$, given by

$$\{R|\mathbf{t}\} = \text{point group operation} R + \text{translation} \, \mathbf{t} \qquad (5.28a)$$

In general

$$\mathbf{t} = \mathbf{T} + \boldsymbol{\tau} \qquad (5.28b)$$

where $\mathbf{T}$ is a primitive translation vector (5.8a) and $\mathbf{t}$ is a non-primitive translation vector (within a primitive cell) and $\boldsymbol{\tau}$ is the increment between these transformations. The operator (5.28a) is called a 'Seitz operator,' and is defined as follows: (a) Choose any point of the lattice as the origin. (b) Perform the point-group operation R about that point. (c) Translate the crystal by $\mathbf{t}$.

The operation of a Seitz operator on a radius vector is given by

$$\{R|\mathbf{t}\}\mathbf{r} = R\mathbf{r} + \mathbf{t} \qquad (5.28c)$$

Note the special cases:

$$\{E|\mathbf{t}\} = \text{pure translation} \qquad (5.28d)$$

$$\{R|0\} = \text{pure point group operation} \qquad (5.28e)$$

$$\{R|\mathbf{t}\}' = \text{'glide-plane' or 'screw' operator} \qquad (5.28f)$$

The latter operations are defined as follows:

Glide-Plane Operation: Translation parallel to a given plane by a non-primitive translation, $\mathbf{t}$, and reflection in a plane containing that vector. This operation brings the crystal structure into coincidence with itself. In this operation, $R = \sigma$ represents reflection. In the glide operation, $\mathbf{t} = \mathbf{T}/2$, as $\sigma^2 = E$, so that performing this operation twice cancels the reflection part of the operation. As shown in Fig. 5.10, in one application of the operation, the effect of $\mathbf{t}$ alone may not carry an atom to an atom. But the effect of twice the glide operation reproduces an atom at location 3 which by definition is displaced from atom 1 by a primitive translation vector.

Note that in the crystal depicted in Fig. 5.10, there are no atoms present in the given crystal at the ends of the two $\mathbf{t}$ vectors. Furthermore, reflection in this operation is through the glide plane.

Screw Operation: Rotation of a lattice about a given axis by $2\pi/n$ ($n = 2, 3, 4, 6$), and translation by $\mathbf{t}$ along that axis. In this operation $R = C_n$ represents rotation. This operation brings the crystal structure into coincidence with itself and is relevant to a crystal composed of parallel helical atomic arrangements.

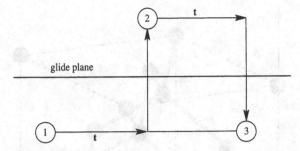

Figure 5.10. Schematic of the glide-plane operation illustrating that for this operation, $2\mathbf{t} = \mathbf{T}$. Note that the glide plane is not a plane of reflection symmetry.

Real Affine Group

The elements $\{R|\mathbf{t}\}$ comprise the *real affine group*, A, with the following group properties:

(a) Closure:

$$\{R|\mathbf{t}\}\{S|\mathbf{t}_1\}\mathbf{r} = \{R|\mathbf{t}\}(S\mathbf{r} + \mathbf{t}_1) = RS\mathbf{r} + R\mathbf{t}_1 + \mathbf{t}$$
$$= \{RS|R\mathbf{t}_1 + \mathbf{t}\}\mathbf{r} \tag{5.29a}$$

So we may write

$$\{R|\mathbf{t}\}\{S|\mathbf{t}_1\} = \{RS|R\mathbf{t}_1 + \mathbf{t}\} = \{RS|\mathbf{t}'\} \in A \tag{5.29b}$$

where it is noted that $R\mathbf{t}_1 + \mathbf{t}$ is a translation.

(b) Inverse:

$$\{R|\mathbf{t}\}^{-1} = \{R^{-1}| - R^{-1}\mathbf{t}\}$$
$$\{R|\mathbf{t}\}\{R^{-1}| - R^{-1}\mathbf{t}\} = \{RR^{-1}|R(-R^{-1}\mathbf{t}) + \mathbf{t}\} = \{E|0\} \tag{5.29c}$$

It follows that elements of A comprise a group.

Space Group

The space group, $\mathcal{G}$, of crystal is obtained from elements of A by restricting translations to primitive translations and point group operations to those relevant to the given crystal. Operations of $\mathcal{G}$ leave the crystal unchanged and are given by the Seitz operators

$$\{R|\mathbf{T}\} \in \mathcal{G} \tag{5.29d}$$

Note that

$$\mathcal{G} \subset A$$

A point group operation on a system maps the system onto itself and leaves one point fixed. A space group operation on a system maps the system onto itself.

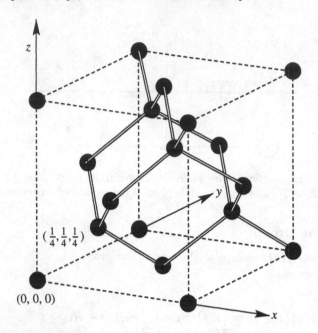

Figure 5.11. Face-centered cubic structure of diamond of the conventional cell with a basis of two carbon atoms. Coordinates of the basis with one atom at the origin are $(0,0,0)$ and $(\frac{1}{4},\frac{1}{4},\frac{1}{4})$. The six face-center atoms have respective coordinates: $(0,\frac{1}{2},\frac{1}{2}),(\frac{1}{2},0,\frac{1}{2}),(\frac{1}{2},\frac{1}{2},0),(1,\frac{1}{2},\frac{1}{2}),(\frac{1}{2},1,\frac{1}{2}),(\frac{1}{2},\frac{1}{2},1)$.

Non-Symmorphic Space Groups; an Example

A *symmorphic group* includes neither glide-plane nor screw operations. Selenium and tellurium crystals are composed of parallel helical atomic arrangements, described by the respective space groups $D_3^4(P3_121)$ and $D_3^6(P3_221)$. (In this notation both Schöenflies and international terms are employed.) The D symbol indicates that both crystals have D_3 symmetries. The superscripts 4, 6 differentiate between the space groups (see also, Section 5.5). The P symbol indicates that both crystals have the hexagon Bravais lattice (Table 5.1). The number 3 in each parentheses indicates that both crystals have threefold rotational (i.e., C_3) symmetry about the principal axis of the helix. The subscript 1 in the space-group symbol for selenium indicates that the rotational symmetry is that of a right-handed screw. The subscript 2 in the space-group symbol for tellurium indicates that rotational symmetry is that of a left-handed screw. The remaining two integers in the space-group symbols indicates that one of the axes normal to the principal axis has C_2 symmetry, and the second axis is void of symmetry. In selenium the screw operation involves a $2\pi/3$ rotation to the right followed by a translation of $a/3$ along the axis of the helix, where a is the lattice constant.

The six elements of the non-symmorphic space group for selenium are

$$D_3^4 = [\{E|0\}, \{C_2|0\}, \{C_3|t\}, \{E|T_1\}, \{E|T_2\}, \{E|T_3\}] \qquad (5.30a)$$

where, we recall, t denotes a non-primitive translation. (Symmorphic space groups are defined in Table 5.3 on page 112.)

Glide Planes and Screw Operations; Diamond Structure

As noted above, glide planes and screw operations do not involve primitive translations but leave the crystal unaltered. They may be illustrated for the diamond structure whose lattice is fcc with two carbon atoms per primitive cell (Fig. 5.11). If one of the two basis atoms is at $(0, 0, 0)$, then the other is at $(\frac{1}{4}, \frac{1}{4}, \frac{1}{4})$. The conventional unit cell of the diamond lattice is a cube with eight atoms per unit cell (with unit edge-lengths) whose positions may be written:

$$(n_1, n_2, n_3), \left(n_1 + \frac{1}{4}, n_2 + \frac{1}{4}, n_3 + \frac{1}{4}\right), \left(n_1, n_2 + \frac{1}{2}, n_3 + \frac{1}{2}\right),$$

$$\left(n_1 + \frac{1}{2}, n_2, n_3 + \frac{1}{2}\right), \left(n_1 + \frac{1}{2}, n_2 + \frac{1}{2}, n_3\right), \left(n_1 + \frac{3}{4}, n_2 + \frac{3}{4}, n_3 + \frac{1}{4}\right),$$

$$\left(n_1 + \frac{1}{4}, n_2 + \frac{3}{4}, n_3 + \frac{3}{4}\right), \left(n_1 + \frac{3}{4}, n_2 + \frac{1}{4}, n_3 + \frac{3}{4}\right),$$

where n_i are positive or negative integers or zero. Whereas the cubic Bravais lattice has O_h symmetry the basis atoms alter this symmetry. For example, an atom originally at $(\frac{1}{4}, \frac{1}{4}, \frac{1}{4})$ under a C_4 rotation (about the z axis), goes to $(-\frac{1}{4}, \frac{1}{4}, \frac{1}{4})$, a point not in the preceding set. However, different symmetries may be found that are not included in the O_h group that describe this crystal. Here are two such operations:

(a) The glide-plane operation. This operation in diamond may be described by the Seitz operator $\{\sigma_{vx}|t\}$ which includes a reflection through the (100) plane followed by a (non-primitive) translation through $[\frac{1}{4}, \frac{1}{4}, \frac{1}{4}]$ (Fig. 5.11). In this operation an atom originally at $(n_1 + \frac{1}{4}, n_2 + \frac{1}{4}, n_3 + \frac{1}{4})$ reflects to $(-n_1 - \frac{1}{4}, n_2 + \frac{1}{4}, n_3 + \frac{1}{4})$ and then translates to $(-n_1, n_2 + \frac{1}{2}, n_3 + \frac{1}{2})$ which is a face center of the original lattice. Note again that in this glide-plane operation, t, is not a primitive translation vector but the crystal is left unaltered.

(b) The screw operation. Consider an axis in diamond parallel to the x axis (Fig. 5.11) that passes through the point $(1, \frac{1}{4}, 0)$ (the screw axis). A C_4 rotation about this axis followed by the translation $[\frac{1}{4}, 0, 0]$ (parallel to the screw axis) maps the crystal onto itself.

In addition to the lattice translations, $\{E|T\}$, the full space group of the diamond crystal includes the 48 operations

$$(E, 8C_3, 3C_2, 6S_4, 6\sigma_d) \quad \text{in the } \{R|0\} \text{ form} \qquad (5.30b)$$

and

$$(i, 8S_6, 3\sigma_v, 6C_4, 6C_2') \quad \text{in the } \{R'|\mathbf{t}\} \text{ form} \qquad (5.30c)$$

where the i operation corresponds to the inversion center $(\frac{1}{8}, \frac{1}{8}, \frac{1}{8})$ and $\mathbf{t} = a(\mathbf{i} + \mathbf{j} + \mathbf{k})/4$ for all R', and $(\mathbf{i}, \mathbf{j}, \mathbf{k})$ are unit vectors parallel to the x, y, z axes, respectively. In product notation, the space group for diamond is labeled $O_h^7 (Fd3m)$ where d indicates a glide plane and the superscript 7 is a space-group number. (Components of this notation are described above for selenium and below for tellurium, Table 5.3.)

Translation and Crystallographic Point Groups

The set of all primitive translations $\{E|\mathbf{T}\}$ comprises the translation group, $\bar{T}$. We note that

$$\bar{T} \subset \mathcal{G} \qquad (5.31a)$$

The set of all elements $\{R|\mathbf{0}\}$ comprises the *Crystallographic Point Group*, and is labeled P_g. In this notation, R is a point group compatible with a Bravais lattice. We note that

$$P_g \subset \mathcal{G} \qquad (5.31b)$$

Let us demonstrate group properties of the translation group. To these ends we introduce the notation [recall (5.8a)]

$$\mathbf{T_n} = \mathbf{a}_1 n_1 + \mathbf{a}_2 n_2 + \mathbf{a}_3 n_3 \qquad (5.31c)$$

where $\mathbf{n} \equiv (n_1, n_2, n_3)$. The inverse is given by

$$\mathbf{T_n^{-1}} = \mathbf{T_{-n}} = \mathbf{a}_1(-n_1) + \mathbf{a}_2(-n_2) + \mathbf{a}_3(-n_3) \qquad (5.31d)$$

The binary operation of the translation group *is addition of* parallel n_i indices. Thus,

$$\mathbf{T_n T_n^{-1}} = \mathbf{T_0}$$

where $\mathbf{T_0}$ is the identity element of the group. It is noted that

$$\mathbf{T_n T_m} = \mathbf{T_m T_n} \qquad (5.31e)$$

It follows that $\bar{T}$ is an Abelian group and that

$$\mathbf{T_n} = \mathbf{T_m T_n T_m^{-1}} \qquad (5.31f)$$

Hence each element of $\bar{T}$ is in a class by itself (Section 2.2, item (d)). Consequently, there are as many one-dimensional irreps as there are elements of the group.

If all symmetries of a crystal are ignored except for translational symmetry, the Bloch functions (5.22c) reduce to the form

$$\varphi_{\mathbf{k}}(\mathbf{r} + \mathbf{T}) = e^{i\mathbf{k} \cdot \mathbf{T}} \varphi_{\mathbf{k}}(\mathbf{r}) \qquad (5.31g)$$

Basis functions of the irreps of $\bar{T}$ are then labeled by the propagation vector $\mathbf{k}$.

Delineation of Space Groups

- There are 73 *symmorphic space groups.*

- There are 157 *nonsymmorphic space groups.*

- There are 230 *space groups.*

- There are 32 *crystallographic point groups.*

Space groups are built up from the crystallographic point groups and translation groups as well as additional symmetries due to atomic basis groupings. Examples of such space groups for the case of the diamond, selenium and tellurium crystals were discussed above.

The 32 P_g groups, together with the corresponding point-group properties (Schönflies symbols, employed in this text) and crystallographic symbols, are listed in Table 5.3. Because only rotations $2\pi/n$ ($n = 2$, 3, 4, 6) are included in these point groups, only point groups associated with these rotational values occur. Note further that groups of the cubic list are subgroups of O_h; groups of the trigonal-hexagonal list are subgroups of D_{6h}. In general, these crystallographic point groups are subgroups of either O_h or D_{6h}. From the character tables listed in Appendix A, it is noted that the maximum degeneracy of any eigenstate with these symmetries is three.

Crystallographic symbols in Table 5.3 are given in international notation. This notation is based on the property that point groups are composed of four basic operations: rotation, reflection, improper rotation and inversion. In this notation:

(1) A number n denotes the presence of an n-fold rotation axis and m denotes σ_d or σ_v reflection planes.

(2) The symbol "$/m$" denotes σ_h reflection planes of a given class that are perpendicular to the symmetry axis.

(3) The symbol $\bar{n}$ denotes an S_n axis.

(4) To distinguish between common symmetry operations in different classes the symbol m is repeated once for every class of planes of mirror symmetry.

For example, in international notation one writes the D_{4h} group as $4/mmm$, indicating a fourfold rotation axis and three classes of reflection symmetry. Symmetry planes of one such class are normal to the symmetry axis. The group $T_d = (E, 8C_3, 3C_2, 6\sigma_d, 6S_4)$ has the international notation $\bar{4}3m$, in which $\bar{4}$ corresponds to the elements S_4, S_4^2 ($\equiv C_2$) and S_4^3; 3 corresponds to C_3; and m corresponds to σ_d planes of a given class.

The Finite Crystal and Periodic Boundary Conditions

In the preceding analysis it is assumed that the crystal being described is of infinite extent. (For example, the translation operator is not defined

Table 5.3

The 32 Crystallographic Point Groups (P_g)

Index	Schönflies Symbol	Crystallographic Symbol	Crystal
1	C_1	1	Triclinic
2	S_2	$\bar{1}$	
3	C_{1h}	m	Monoclinic
4	C_2	2	
5	C_{2h}	$2/m$	
6	C_{2v}	$mm2$	Orthorhombic
7	D_2	222	
8	D_{2h}	mmm	
9	C_4	4	Tetragonal
10	S_4	$\bar{4}$	
11	C_{4h}	$4/m$	
12	C_{4v}	$4mm$	
13	D_{2d}	$\bar{4}2m$	
14	D_4	422	
15	D_{4h}	$4/mmm$	
16	C_3	3	Rhombohedral
17	S_6	$\bar{3}$	
18	C_{3v}	$3m$	
19	D_3	32	
20	D_{3d}	$\bar{3}m$	
21	C_{3h}	$\bar{6}$	Hexagonal
22	C_6	6	
23	C_{6h}	$6/m$	
24	D_{3h}	$\bar{6}m2$	
25	C_{6v}	$6mm$	
26	D_6	622	
27	D_{6h}	$6/mmm$	
28	T	23	Cubic
29	T_h	$m3$	
30	T_d	$\bar{4}3m$	
31	O	432	
32	O_h	$m3m$	

for a finite crystal.) To apply this analysis to a finite crystal we employ the assumption of periodic boundary conditions. Let the given crystal have lattice constants (a, b, c) in the three Cartesian directions and dimensions $(2Na, 2Nb, 2Nc; N \gg 1$, and integer), symmetrically oriented about the three Cartesian axes. At the boundary plane $x = Na$ imagine that the crystal turns on itself so that this plane meets the crystal $x = -Na$, such that any crystal property at the $x = Na$ plane matches the crystal property at $x = -Na$. Repeating this process for the remaining two dimensions offers a method of applying the theory of space groups to a finite crystal.

Factor-Group Theorem

Let us show that the translation group $\bar{T}$ is an invariant subgroup of the space group $\mathcal{G}$. That is,

$$\mathcal{G}\bar{T}\mathcal{G}^{-1} = \bar{T} \tag{5.32a}$$

or, equivalently (Problem 5.11),

$$\{R|\mathbf{T}\}\{E|\mathbf{T}\}\{R|\mathbf{T}\}^{-1} = \{E|\mathbf{T}'\} \tag{5.32b}$$

where $\mathbf{T}'$ is a primitive translation vector. This invariance property indicates that one may form the factor group, $\mathcal{G}/\bar{T}$, which is composed of the complexes

$$\mathcal{G}/\bar{T} = \{\bar{T}, \mathcal{G}_1'\bar{T}, \mathcal{G}_2'\bar{T}, \ldots\} \tag{5.32c}$$

where, in accord with Section 5.1,

$$\mathcal{G}_1' = \{R_i|0\} \in \mathcal{G} - \bar{T} \tag{5.32d}$$

With (5.6c) we note that the irreps of $\mathcal{G}/\bar{T}$ correspond to irreps of $\mathcal{G}$. Note that $\mathcal{G}_1'$ elements are included in P_g.

A fundamental theorem of crystal physics is that $\mathcal{G}/\bar{T}$ is isomorphic to P_g. It follows that irreps of P_g correspond to the irreps of $\mathcal{G}$. Let us demonstrate the isomorphism between $\mathcal{G}/\bar{T}$ and P_g for symmorphic point groups. If elements of P_g are $R, Q, \ldots$, then components of the factor group $\mathcal{G}/\bar{T}$ are given by

$$\{E|\bar{T}\}, [\{E|\bar{T}\}\{R|0\}], [\{E|\bar{T}\}\{Q|0\}], \ldots, C_R, C_Q \ldots \tag{5.33a}$$

where C_R is the complex obtained by multiplying all elements of $\{E|\bar{T}\}$ with $\{R|0\}$, etc. The square brackets in (5.33a) indicate that there are as many entries in each bracket as there are elements of the group $\bar{T}$. For the demonstration at hand consider the multiplication of the two complexes (with $\{E|\bar{T}\}\{R|0\} = \{R|\bar{T}\}$)

$$C_R C_q = [\{R|\bar{T}\}][\{Q|\bar{T}'\}] = [\{RQ|Q\bar{T}+\bar{T}'\}] = [\{RQ|\bar{T}''\}] = [\{V|\bar{T}''\} = C_V \tag{5.33b}$$

It follows that the factor-group product

$$C_R C_Q = C_V \tag{5.33c}$$

corresponds to the point-group product

$$RQ = V \qquad (5.33d)$$

We conclude that the factor group $\mathcal{G}/\bar{T}$ and the crystallographic point group P_g have the same multiplication tables. This establishes the isomorphism between the two groups.

The observation that irreps of $\bar{T}$ and P_g, respectively, correspond to irreps of $\mathcal{G}$ is consistent with the known theorem that all the irreps of $\mathcal{G}$ may be expressed in terms of the irreps of $\bar{T}$ and the irreps of P_g.[2]

Brillouin Zone Revisited

In Section 5.3 the Brillouin zone was described in terms of surfaces of constructive interference in reciprocal lattice space, for X-ray scattering. Related properties of the Brillouin zone come into play in study of the propagating electron Bloch waves (5.22c) in a crystal. Eigenenergies corresponding to Bloch waves are found to yield a band structure. This property is clearly exhibited by an E-k diagram for one-dimensional motion. For free-particle motion the E-k relation is the parabola

$$E = \frac{\hbar^2 k^2}{2m} \qquad (5.34a)$$

For propagation in a periodic structure with lattice constant a, this parabola has discontinuities at values of k corresponding to the edges of the one-dimensional Brillouin zone, $ka = n\pi$, where n is a positive or negative integer. See Fig. 5.12. The first Brillouin zone corresponds to the k values

$$-\pi \leq ka \leq \pi \qquad (5.34b)$$

The second Brillouin zone corresponds to the k values

$$\pi \leq |ka| \leq 2\pi \qquad (5.34c)$$

Recalling (5.26) indicates that $E(k)$ is periodic in k, with period given by the *rlv*. In the one-dimensional problem G has lengths $n\pi/a$. This periodicity of the $E(k)$ curves is such that on each branch

$$E^{(j)}(k) = E^{(j)}\left(k + \frac{n\pi}{a}\right) \qquad (5.34d)$$

where n is an even integer and the j superscript labels the corresponding branch. The graph containing sections of all such curves in the interval $|ka| \leq \pi$ is called the 'reduced zone picture' (Fig. 5.12(c)).

Eigenenergies of an electron propagating through a crystal comprise energy bands separated by energy gaps. The first energy gap of a band

---

[2]For further discussion, see J.F. Cornwell, *Group Theory in Physics* (see Bibliography).

structure is shown in Fig. 5.12(b) and the reduced-zone picture is shown in Fig. 5.12(c).

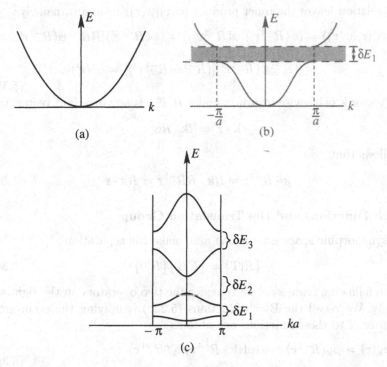

Figure 5.12. (a) The free-particle E-k parabola. (b) Related diagram for propagation in a periodic array, illustrating discontinuities at the edge of the first Brillouin zone. The first, and part of the second, energy bands are shown. The first energy gap is labeled δE_1. (c) Band structure in the reduced-zone picture showing the first four energy bands and related energy gaps.

In a metal, energy states in the highest band (the 'conduction' band) are partially occupied, and the material conducts. If states in the conduction band are entirely unoccupied, and the energy gap between the bottom of the conduction band and the top of the next lower band in energy (the 'valence' band) is relatively large, the crystal is an insulator. For example the energy gap in diamond is approximately 5.5 eV. In a semiconductor, the energy gap is relatively narrow, so that, for example, the crystal will conduct under thermal agitation. In the semiconductor GaAs the energy gap at 0 K is approximately 1.17 eV.

Symmetry of the Wavefunction

Consider first the operation of a point-group operator R on a function $\psi(\mathbf{r})$. We note that R operates on coordinates such that $\mathbf{r}' = R\mathbf{r}$ and $\mathbf{r}' \cdot \mathbf{r}' = \mathbf{r} \cdot \mathbf{r}$.

The correct relation for the operation of R on $\psi(\mathbf{r})$ is

$$R\psi(\mathbf{r}) = \psi(R^{-1}\mathbf{r}) \tag{5.35a}$$

This relation leaves the inner product $\langle\varphi(\mathbf{r})|\varphi(\mathbf{r})\rangle$ invariant, namely,

$$\langle\psi(\mathbf{r})|\psi(\mathbf{r})\rangle = \langle\psi(R^{-1}\mathbf{r}')|\psi(R^{-1}\mathbf{r}')\rangle = \langle\psi(R^{-1}\mathbf{r}')|RR^{-1}\psi(R^{-1}\mathbf{r}')\rangle$$

$$= \langle(R^{-1}\psi(R^{-1}\mathbf{r}')|(R^{-1}\psi(R^{-1}\mathbf{r}')\rangle = \langle\psi(\mathbf{r}')|\psi(\mathbf{r}')\rangle \tag{5.35b}$$

The vector $\mathbf{k}$ is likewise invariant under R. If $\mathbf{z}$ is any invariant vector, then

$$\mathbf{k} \cdot \mathbf{z} = R\mathbf{k} \cdot R\mathbf{z}$$

It follows that

$$\mathbf{k} \cdot R^{-1}\mathbf{r} = R\mathbf{k} \cdot RR^{-1}\mathbf{r} = R\mathbf{k} \cdot \mathbf{r} \tag{5.35c}$$

Bloch Functions and the Translation Group

For symmorphic space groups one may make the separation

$$\{R|\mathbf{T}\} = \{E|\mathbf{T}\}\{R|0\} \tag{5.36a}$$

In the following discussion we consider the two operators on the right separately. We recall the Bloch functions (5.22c). Applying the point-group operator R to these functions one obtains

$$R\varphi_\mathbf{k}(\mathbf{r}) = \varphi_\mathbf{k}(R^{-1}\mathbf{r}) = \exp(i\mathbf{k} \cdot R^{-1}\mathbf{r})u_\mathbf{k}(R^{-1}\mathbf{r})$$

$$= \exp(iR\mathbf{k} \cdot \mathbf{r})u_\mathbf{k}(R^{-1}\mathbf{r}) \equiv \exp(iR\mathbf{k} \cdot \mathbf{r})u_{R\mathbf{k}}(\mathbf{r}) \equiv \varphi_{R\mathbf{k}}(\mathbf{r}) \tag{5.36b}$$

so that

$$R\varphi_\mathbf{k}(\mathbf{r}) = \varphi_{R\mathbf{k}}(\mathbf{r}) \tag{5.36c}$$

So the effect of operating on $\varphi_\mathbf{k}(\mathbf{r})$ with R is to produce an eigenfunction also in Bloch form but with $\mathbf{k}$ transformed to $R\mathbf{k}$, corresponding to the same eigenenergy as $\varphi_\mathbf{k}(\mathbf{r})$.

We wish to show that these Bloch functions are a basis for the translation group, $\bar{T}$. Consider the form

$$\{E|\mathbf{T}\}\varphi_{R\mathbf{k}} = \{E|\mathbf{T}\}\varphi_\mathbf{k}(R^{-1}\mathbf{r})$$

$$= \varphi_\mathbf{k}(\{E|\mathbf{T}\}^{-1}R^{-1}\mathbf{r}) = \varphi_\mathbf{k}(R^{-1}\mathbf{r} - R^{-1}\mathbf{T})$$

$$= \exp i\mathbf{k} \cdot [R^{-1}\mathbf{r} - \mathbf{T}]u_\mathbf{k}(R^{-1}\mathbf{r} - \mathbf{T}) \tag{5.36d}$$

$$= \exp i\mathbf{k} \cdot [R^{-1}\mathbf{r} - \mathbf{T}]u_\mathbf{k}(R^{-1}\mathbf{r}) = \exp i\mathbf{k} \cdot [R^{-1}\mathbf{r} - \mathbf{T}]u_{R\mathbf{k}}(\mathbf{r}) \tag{5.36e}$$

$$= e^{-i\mathbf{k}\cdot\mathbf{T}}e^{iR\mathbf{r}\cdot\mathbf{k}}u_{R\mathbf{k}}(\mathbf{r}) = e^{-i\mathbf{k}\cdot\mathbf{T}}\varphi_{R\mathbf{k}}(\mathbf{r}) \tag{5.36f}$$

In (5.36d) we noted that $R^{-1}\mathbf{T}$ is some lattice translation, $\mathbf{T}'$ (labeled $\mathbf{T}$). In (5.36e) the periodicity of $u(\mathbf{r})$, (5.23c), was recalled. In (5.36g), (5.35c) was used. Equating the first term in the preceding equations to the right side of (5.36f) gives

$$\{E|\mathbf{T}\}\varphi_{R\mathbf{k}}(\mathbf{r}) = e^{-i\mathbf{k}\cdot\mathbf{T}}\varphi_{R\mathbf{k}}(\mathbf{r}) \qquad (5.36g)$$

which indicates that the function $\varphi_{R\mathbf{k}}(\mathbf{r})$ may be taken as a basis function of the $R\mathbf{k}$th irrep of the translation group.

We wish to obtain the Schrödinger equation for $u_{R\mathbf{k}}(\mathbf{r})$. First we note the relations

$$\varphi_{\mathbf{k}}(R^{-1}\mathbf{r}) \equiv \varphi_{R\mathbf{k}}(\mathbf{r}) = e^{iR\mathbf{r}\cdot\mathbf{k}}u_{R\mathbf{k}}(\mathbf{r}) \qquad (5.37a)$$

$$u_{R\mathbf{k}}(\mathbf{r}) \equiv u_{\mathbf{k}}(R^{-1}\mathbf{r}) \qquad (5.37b)$$

[The first of these follows from (5.36g).] Recalling (5.22a),

$$H\varphi_{\mathbf{k}}(\mathbf{r}) = E(\mathbf{k})\varphi_{\mathbf{k}}(\mathbf{r}) \qquad (5.37c)$$

and operating on this equation with $\{R|0\}$ gives

$$H\varphi_{\mathbf{k}}(R^{-1}\mathbf{r}) = E(\mathbf{k})\varphi_{\mathbf{k}}(R^{-1}\mathbf{r}) \qquad (5.37d)$$

which, with (5.37a), gives

$$He^{iR\mathbf{r}\cdot\mathbf{k}}u_{R\mathbf{k}}(\mathbf{r}) = E(\mathbf{k})e^{iR\mathbf{r}\cdot\mathbf{k}}u_{R\mathbf{k}}(\mathbf{r}) \qquad (5.37e)$$

Repeating operations leading to (5.23a), with (5.36c), gives the Schrödinger equation

$$\left[-\frac{\hbar^2\nabla^2}{2m} + \frac{\hbar}{m}R\mathbf{k}\cdot\mathbf{p} + V(\mathbf{r})\right]u_{R\mathbf{k}}(\mathbf{r}) = \left[E(\mathbf{k}) - \frac{\hbar^2 k^2}{2m}\right]u_{R\mathbf{k}}(\mathbf{r}) \qquad (5.37f)$$

which is the generalization of (5.23a) for the Bloch function $u_{R\mathbf{k}}(\mathbf{r})$.

The Group of k and the Star of k. General Points

We turn next to the $\{R|0\}$ components of $\mathcal{G}$, that is, the crystallographic point group, P_g. The 'group of $\mathbf{k}$' is the subgroup of P_g that leaves $\mathbf{k}$ invariant. Elements $(R_{\mathbf{k}})$ of this subgroup have the property

$$R_{\mathbf{k}}\mathbf{k} = \mathbf{k} + \mathbf{G} \qquad (5.38a)$$

[Recall (5.19).] This subgroup is labeled $P(\mathbf{k})$ so that

$$R_{\mathbf{k}} \in P(\mathbf{k}) \subset P_g \qquad (5.38b)$$

For the square lattice in two dimensions, point symmetries are included in the C_{4v} group with elements $(E, 2C_4, C_2, 2\sigma_v, 2\sigma_d)$. If one operates successively on a given $\mathbf{k}$ vector (within the first Brillouin zone) with $R \in C_{4v}$, one obtains what is known as a *star of* $\mathbf{k}$. If $\mathbf{k}$ is chosen so that it does not terminate on the Brillouin zone, nor lie along any symmetry axis, then it

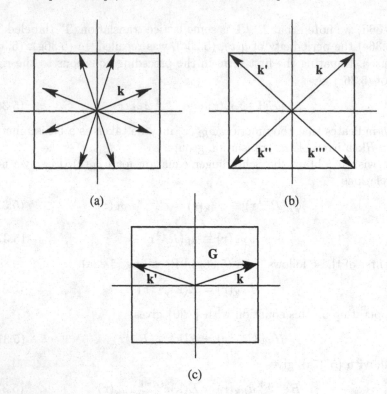

Figure 5.13. Transformations of **k** in the first Brillouin zone of a two-dimensional square lattice. In (a) **k** is a 'general point,' for which $\{E|0\}$ is the only element of P_g which leaves **k** invariant. Orientations of **k** stem from the σ_v and σ_d reflections. In (b), (c), $h[P(\mathbf{k})] > 1$ and in each case **k** is a 'special point.' In both these latter cases **k** is related to equivalent primed **k** vectors through a reciprocal lattice **G** vector.

is called a *general point*. In this event each operation of R carries **k** into a new vector. If there are h such vectors, the star will have h 'points' (Fig. 5.13(a)). The eigenstates corresponding to the $h\mathbf{k}$ values are distinct but, with (5.37), correspond to the same eigenenergy. These h eigenfunctions, $\varphi_{R\mathbf{k}}(\mathbf{r})$, form the basis of an h-dimensional irrep of $\mathcal{G}$. However, for this case, the group of **k** has one element, $\{E|0\}$, and the irrep is one-dimensional, and point symmetries add nothing to the translational symmetry operations. Irreps and basis functions of $\mathcal{G}$ for this case are determined by the translational symmetry of the crystal.

The Γ Point

Note that at the Γ point,

$$R_{\mathbf{k}}0 = 0 + \mathbf{G} = \mathbf{G} \qquad (5.38c)$$

where the **k** vector $\mathbf{0} \equiv (0,0,0)$. It follows that the group of **k** at the Γ point is composed of the set of all **G** vectors. As noted previously, the set of **G** vectors generates reciprocal lattice space. As direct and reciprocal lattice spaces are described by the same point group, it follows that the group of **k** at the Γ is isomorphic to the point group of the direct lattice (Problem 5.19).

Special Points

Consider the **k** vector shown in Fig. 5.13(b), which touches the zone boundary. Operating on this vector with σ_v transforms it into **k**'. But for this vector

$$\mathbf{k}' = \mathbf{k} + \mathbf{G} \qquad (5.38d)$$

so that **k** and **k**' are equivalent and we may say that **k** is invariant under the operation σ_v. The same invariance applies for all elements of C_{4v} so that for this **k** vector, $P(\mathbf{k}) = C_{4v}$, whose operations on **k** are shown as higher primed **k** vectors in Fig. 5.13(b). The star of **k** for this example contains one vector, **k**. For the **k** vector shown in Fig. 5.13(c), σ_v operating on **k** is described (5.38c). However, for this case no other element of C_{4v} leaves **k** invariant and we conclude that $P(\mathbf{k}) = (E, \sigma_v)$. Again the star of **k** contains only one vector, **k**. In situations for which **k** gives rise to $P(\mathbf{k})$ with more than one element, **k** is called a *special point*.

To summarize (for symmorphic groups)

Star of **k**: for **k** a *general point*.

- **k** does not terminate on the zone boundary nor lie along any symmetry axis.

- Star composed of vectors: $R\mathbf{k}$, for all $\{R|0\} \in P_g$.

- Corresponding to h vectors in a star of **k** there exist h functions, $\{\varphi_{R\mathbf{k}}(\mathbf{r})\}$, which comprise a basis for an h-dimensional irrep of $\mathcal{G}$.

- Elements of $P(\mathbf{k})$: $\{E|0\}$

Star of **k**: for **k** a *special point* ($R_k\mathbf{k} = \mathbf{k} + \mathbf{G}$).

- Elements of $P(\mathbf{k})$: $\{E|0\}, \{R_1|0\}, \{R_2|0\}, \dots; h[P(\mathbf{k})] > 1$.

- $P(\mathbf{k}) \subset P_g$. Irreps of $P(\mathbf{k})$ are called the 'small representation.'

If the number of vectors in the star of **k** (for **k** a special point) is $h(\mathbf{k})$ and $h[P(\mathbf{k})]$ is the order of the group of **k**, then

$$h(P_g) = h(\mathbf{k})h[P(\mathbf{k})] \qquad (5.38e)$$

That is, for every **k** vector in the star of **k**, there are $h[P(\mathbf{k})]$ elements that are invariant and contribute to the point group P_g.

5.5   Application to Semiconductor Materials

As described briefly above Fig. 5.12, conductivity of materials depends on the occupancy of the higher lying energy bands of the crystal. A semiconductor has the property of having a relatively small energy gap between valence and conduction bands. Thus, whereas the conduction band of a semiconductor is empty at 0 K temperature, at higher temperatures thermal agitation causes the material to conduct. For example, in silicon the gap is 1.11 eV and in germanium it is 0.72 eV (compared to the insulator, diamond, which has an energy gap of 6 eV).

The majority of semiconducting materials crystallize into one of four structures: diamond (O_h^7); zinc blend or sphalerite (T_d^2); wurtzite (C_6^4); rocksalt (O_h^5). In these group citations, the superscripts refer to the following. As there are 230 space groups and only 32 crystallographic point groups, a means of identifying which space group a given point group belongs to is required. The superscripts serve this purpose. (This is the so-called Schönflies notation.) In this notation a superscript 1 always indicates a crystal described by a symmorphic group.[3]

The point group of symmetry for a sphalerite crystal is the T_d group. Let us obtain the group of k for a lattice with T_d symmetry (see Fig. 2.4). We recall that this group has elements $(E, 8C_3, 3C_2, 6S_4, 6\sigma_d)$. The group of k in this instance is the subgroup of T_d which, when acting on k, leaves it invariant or changes it by a reciprocal lattice vector. For k values in the direction [100] there are three related groups:

(a) At the Γ point, the group of k is isomorphic to the T_d group.

(b) At the special point $\mathbf{k} = 2\pi/a(1, 0, 0)$, labeled the X point (Figs. 5.9(a), 5.14), the operations $(E, C_2, 2\sigma)$ leave k at the X point. The operations $2S_4$ and $2C_2$ transform X to $-X$ which are separated by the **G** vector $2\pi/a(2, 0, 0)$. It follows that the group of k for this special point is the D_{2d} group.

(c) At the general point $k_x < G$, at displacements Δ, the group of k has only four elements $(E, C_2, 2\sigma)$ which comprise the group C_{2v}. Note that in this case, $2S_4$ and $2C_2$ transform k_x to $-k_x$. These vectors are not separated by a **G** vector and do not contribute to the group of k.

These results are summarized in the character tables listed in Table 5.4. Notation familiar to solid-state physics is included in this table in which the Γ point refers to the value $\mathbf{k} = 0$ (see Fig. 5.9) as well as to chemical notation employed in the present test. (For further discussion see Koster (1957), Bibliography.)

---

[3]This scheme is described in N.F. Henry and K. Lonsdale; G. Burns (see Bibliography).

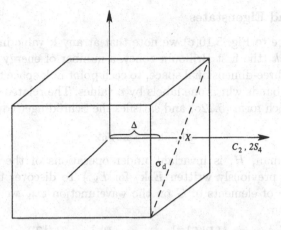

Figure 5.14. Operations on a cube in k-space contributing to the group of **k** at the X point for the T_d cubic group. The origin of the cube is the Γ point. The edge-length of the cube has the length of twice a reciprocal lattice vector, $G = 2\pi/a$. The first Brillouin zone of a T_d crystal is contained in the cube.

Table 5.4

Character Tables for Three Groups of k for k Vectors Along the Direction $[1,0,0]$ Through the X Point

		T_d	E	$8C_3$	$6S_4$	$3C_2$	$6\sigma_d$	
Γ_1	Γ_1	A_1	1	1	1	1	1	
Γ_2	Γ_2	A_2	1	1	-1	1	-1	
Γ_{12}	Γ_3	E	2	-1	0	2	0	(a)
Γ_{25}	Γ_5	T_1	3	0	1	-1	-1	$k = 0$
Γ_{15}	Γ_4	T_2	3	0	-1	-1	1	

	D_{2d}	E	$2S_4$	C_2	$2C_2'$	$2\sigma_d$	
X_1	A_1	1	1	1	1	1	
X_4	A_2	1	1	1	-1	-1	
X_2	B_1	1	-1	1	1	-1	(b)
X_3	B_2	1	-1	1	-1	1	X point
X_5	E	2	0	-2	0	0	

	C_{2v}	E	C_2	σ_v	σ_v'	
Δ_1	A_1	1	1	1	1	
Δ_2	A_2	1	1	-1	-1	(c)
Δ_3	B_1	1	-1	1	-1	Δ axis
Δ_4	B_2	1	-1	-1	1	

Energy Band Eigenstates

With reference to Fig. 5.10(c) we note that at any $\mathbf{k}$ value in the domain $-ak \leq \pi \leq ak$ (the first Brillouin zone), a number of energy bands exist. Likewise, in three-dimensional space, to each point in $\mathbf{k}$ space there exists a set of energy bands which one labels by n values. The related wavefuction, $\varphi_{n\mathbf{k}}$, is in Bloch form (5.22c) and satisfies the Schrödinger equation

$$H\varphi_{n\mathbf{k}} = E_{n\mathbf{k}}\varphi_{n\mathbf{k}} \tag{5.39a}$$

The Hamiltonian, H, is invariant under operations of the space group G. (We have previously written $E(\mathbf{k})$ for $E_{\mathbf{k}}$.) To discover the effects of the operation of elements of G on the wavefunction $\varphi_{n\mathbf{k}}$ we consider the transformation

$$\{R|\mathbf{T}\}\varphi_{n\mathbf{k}}(\mathbf{r}) = \varphi_{n\mathbf{k}}(\{R|\mathbf{T}\}^{-1}\mathbf{r}) = \varphi_{n\mathbf{k}}(R^{-1}\mathbf{r} - R^{-1}\mathbf{T})$$

$$= u_{n\mathbf{k}}(R^{-1}\mathbf{r} - R^{-1}\mathbf{T})\exp[-i\mathbf{k} \cdot (R^{-1}\mathbf{r} - R^{-1}\mathbf{T})] \tag{5.39b}$$

With (5.35c) and noting that $R^{-1}\mathbf{T} \equiv \mathbf{T}'$ is a translation vector, and $R^{-1}\mathbf{r} \equiv \mathbf{r}'$ is a point in the lattice, one obtains

$$\{R|\mathbf{T}\}\varphi_{n\mathbf{k}}(\mathbf{r}) = u_{n\mathbf{k}}(\mathbf{r}' - \mathbf{T})\exp[-i\mathbf{k} \cdot (R\mathbf{r} - \mathbf{T}')] = [u_{n\mathbf{k}}e^{i\mathbf{k}\cdot\mathbf{T}'}]\exp(-iR\mathbf{k}\cdot\mathbf{r}) \tag{5.39c}$$

The function in square brackets has the translational periodicity of the lattice so that operation on the wavefunction by $\{R|\mathbf{T}\}$ gives another wavefunction of the Bloch form associated with the $\mathbf{k}$ vector, $R\mathbf{k}$ [cf., (5.36c)]. We may conclude that

$$E_n(\mathbf{k}) = E_n(R\mathbf{k}) \tag{5.39d}$$

This relation implies that an eigenenergy in a given band, as a function of $\mathbf{k}$, possesses the symmetry of the crystal lattice which, in one dimension, assumes the form (5.34d).

Irreps of G. The Space Group of k

Consider a degenerate set of eigenstates of the crystal Hamiltonian with symmetries of the space group $\mathcal{G}$:

$$\varphi_{n_1\mathbf{k}_1}(\mathbf{r}), \varphi_{n_2\mathbf{k}_2}(\mathbf{r}), \cdots \tag{5.40a}$$

This set forms the basis of some irrep of $\mathcal{G}$ which we label $\Gamma^\mathcal{G}$. That is,

$$\{R|\mathbf{T}\}\varphi_{n_i\mathbf{k}_i} = \Sigma_j \varphi_{n_j\mathbf{k}_j}\Gamma^\mathcal{G}_{ij}(\{R|\mathbf{T}\}) \tag{5.40b}$$

where Γ_{ij} are matrix elements of the given irrep.

The subgroup of $\mathcal{G}$ composed of elements $\{R_k|\mathbf{T}\}$ is called the *space group of* $\mathbf{k}$ and labeled $\mathcal{G}_{\mathbf{k}}$, so that

$$\{R_{\mathbf{k}}|\mathbf{T}\} \in \mathcal{G}_{\mathbf{k}} \tag{5.41a}$$

The functions $\{\varphi_{n\mathbf{k}}\}$ are a basis for this subgroup, namely,

$$\{R_\mathbf{k}|\mathbf{T}\}\varphi_{n_i\mathbf{k}} = \Sigma_m\varphi_{n_m\mathbf{k}}\Gamma_{mj}^{\mathcal{G}_\mathbf{k}}(\{R_\mathbf{k}|\mathbf{T}\}) \tag{5.41b}$$

We wish to establish a relation between irreps of $\mathcal{G}_\mathbf{k}$ and the group of $\mathbf{k}$, $P(\mathbf{k})$, for symmorphic groups. Again, developing the right side of (5.41b) for this class of groups gives

$$\{R_\mathbf{k}|\mathbf{T}\}\varphi_{n_j\mathbf{k}} = \{E|\mathbf{T}\}\{R_\mathbf{k}|\mathbf{0}\}\varphi_{n_j\mathbf{k}} = \{E|\mathbf{T}\}\Sigma_m\varphi_{n_m\mathbf{k}}\Gamma_{mj}^{\mathcal{G}_\mathbf{k}}(\{R_\mathbf{k}|\mathbf{0}\})$$

$$= \Sigma_m\varphi_{n_m\mathbf{k}}e^{-i\mathbf{k}\cdot\mathbf{T}}\Gamma_{mj}^{\mathcal{G}_\mathbf{k}}(\{R_\mathbf{k}|\mathbf{0}\})$$
$$\tag{5.41c}$$

In the last step we recalled (5.39c) with $R = E$ and (5.40b) with $\mathbf{T} = 0$. Comparing the last equation with (5.41b) indicates that

$$\Gamma^{\mathcal{G}_k}(\{R_\mathbf{k}|\mathbf{T}\}) = e^{-i\mathbf{k}\cdot\mathbf{T}}\Gamma^{\mathcal{G}_k}(\{R_\mathbf{k}|\mathbf{0}\}) \tag{5.41d}$$

The irreps of the space group $\{R_k|\mathbf{T}\}$ and those of the group of $\mathbf{k}$, $P(\mathbf{k})$, are equal except for a phase factor that is independent of spatial coordinates. Thus, the basis eigenfunctions $\{\varphi_{n\mathbf{k}}\}$ of irreps of the space group of $\mathbf{k}$, $\mathcal{G}_\mathbf{k}$ are likewise basis functions of the irreps of the space group of $\mathbf{k}$, $\mathcal{G}_\mathbf{k}$ (5.40c) and the irreps of the point group of $\mathbf{k}$, $P(\mathbf{k})$. Furthermore, the basis eigenfunctions $\{\varphi_{n\mathbf{k}}\}$ comprise a basis for this irrep, namely,

$$\{R_\mathbf{k}|\mathbf{0}\}\varphi_{n\mathbf{k}} = \Sigma_m\varphi_{n_m}\Gamma_{nm}^{\mathcal{G}_\mathbf{k}}(\{R_\mathbf{k}|\mathbf{0}\}) \tag{5.41e}$$

Table 5.5 lists the space groups and subgroups introduced in this chapter.

Table 5.5
Crystallographic Groups

Group	Name	Element
A	Real affine group	$\{R|\mathbf{t}\}$
$\mathcal{G}$	Space group	$\{R|\mathbf{T}\}$
P_g	Crystallographic point group	$\{R|\mathbf{0}\}$
$\bar{T}$	Translation group	$\{E|\mathbf{T}\}$
$\mathcal{G}_k$	Space group of $\mathbf{k}$	$\{R_\mathbf{k}|\mathbf{T}\}$
$P(\mathbf{k})$	Group of $\mathbf{k}$	$\{R_\mathbf{k}|\mathbf{0}\}$

5.6 Time Reversal, Space Inversion and Double Space Groups

We consider first, properties of electron band-energies of a crystal that obey: (a) time-reversal symmetry, (b) space-inversion symmetry, or (c) time-reversal and space-inversion symmetries. Double space groups are then introduced relevant to spin-orbit coupling and applied to change in degeneracies of silicon and germanium with this interaction turned on. The section concludes with a description of the irreps of the O_h double group.

We recall that the time-reversal operation in quantum mechanics involves two processes: $t \to -t$, and complex conjugation of the wave function and its Hamiltonian. If the Hamiltonian is spin dependent, then this operation includes the spin reversal, $\alpha \to \beta, \beta \to \alpha$, where (α, β) are spin (up, down) states of an electron. Consider the Schrödinger equation for a Bloch wavefunction with the spin state α,

$$H\varphi_{\mathbf{k}\alpha} = E(\mathbf{k}, \alpha)\, \varphi_{\mathbf{k}\alpha} \tag{5.42a}$$

where the spin-dependent Bloch wavefunction is given by

$$\varphi_{\mathbf{k}\alpha} = e^{i\mathbf{k}\cdot\mathbf{r}} u_{\mathbf{k}}(\mathbf{k})\alpha \tag{5.42b}$$

Let us call the *time-reversal operator* Ξ. If the crystal possesses time-reversal symmetry then

$$[H, \Xi] = 0 \tag{5.43a}$$

so that $\varphi_{\mathbf{k}\alpha}$ and $\Xi\varphi_{\mathbf{k}\alpha}$ correspond to the same eigenenergy. We note that

$$\Xi\varphi_{\mathbf{k}\alpha} = e^{-i\mathbf{k}\cdot\mathbf{r}} u_{\mathbf{k}}^*(\mathbf{r})\beta = \varphi_{-\mathbf{k}\beta} \tag{5.43b}$$

where we noted that $u_{\mathbf{k}}^*(r) = u_{-\mathbf{k}}(r)$. The eigenfunction on the right of (5.43b) satisfies the Schrödinger equation

$$H\varphi_{-\mathbf{k}\beta} = E(-\mathbf{k}, \beta)\varphi_{-\mathbf{k}\beta} \tag{5.44}$$

As $\varphi_{\mathbf{k}\alpha}$ and $\Xi\varphi_{\mathbf{k}\alpha}$ correspond to the same eigenenergy, it follows that:

Time-reversal symmetry: $E(\mathbf{k}, \alpha) = E(-\mathbf{k}, \beta)$

This relation is valid independent of the spatial symmetries of the crystal.

Consider next that a crystal is invariant under space inversion. For clarity of presentation we label the space-inversion operator Λ (previously labeled i), so that for the given crystal

$$[H, \Lambda] = 0 \tag{5.45a}$$

Again, this relation implies that $\varphi_{\mathbf{k}\alpha}$ and $\Lambda\varphi_{\mathbf{k}\alpha}$ correspond to the same eigenenergy. We obtain

$$\Lambda\varphi_{\mathbf{k}\alpha} = e^{-i\mathbf{k}\cdot\mathbf{r}} u_{\mathbf{k}}(-\mathbf{r})\alpha \tag{5.45b}$$

If the crystal is invariant to Λ, then stemming from the equation of motion (5.23a) for $u_{\mathbf{k}}(\mathbf{r})$, one finds $u_{\mathbf{k}}(-\mathbf{r}) = u_{-\mathbf{k}}(\mathbf{r})$ so that

$$\Lambda\varphi_{\mathbf{k}\alpha} = \varphi_{-\mathbf{k}\alpha} \qquad (5.45c)$$

This eigenfunction satisfies the Schrödinger equation

$$H\varphi_{-\mathbf{k}\alpha} = E(-\mathbf{k}, \alpha)\varphi_{-\mathbf{k}\alpha} \qquad (5.45d)$$

Note that space inversion does not affect spin. There results:

Space-inversion symmetry: $E(\mathbf{k}, \alpha) = E(-\mathbf{k}, \alpha)$

This degeneracy is removed if the crystal has no inversion center, such as is the case for the InSb crystal. For a crystal with both time-reversal and space-inversion symmetries, the relations (5.45) and (5.47) give:

Time-reversal and space-inversion symmetry: $E(\mathbf{k}) = E(-\mathbf{k})$

These results are summarized in Table 5.6.

Table 5.6

Crystal Symmetry Eigenenergy Relations

Time-Reversal Symmetry	Space-Inversion Symmetry	Time-Reversal and Space-Inversion Symmetry
$E(\mathbf{k}, \alpha) = E(-\mathbf{k}, \beta)$	$E(\mathbf{k}, \alpha) = E(-\mathbf{k}, \alpha)$	$E(\mathbf{k}) = E(-\mathbf{k})$

This table indicates that for crystals with space-inversion symmetry all bands are at least twofold degenerate. At $k = 0$ (the Γ point) condition (5.45) indicates that electrons with spin up or spin down have equal energies, regardless of inversion symmetry, again inferring energy bands with at least twofold symmetry.

Double Space Groups

Double space groups come into play when electron spin functions are introduced. This is due to the following. Consider the character of the ℓth irrep of $O(3)^+$, given by (4.19). Under a rotation through $\alpha + 2\pi$ one obtains

$$\chi^\ell(\alpha + 2\pi) = \frac{\sin[(\ell + \frac{1}{2})(\alpha + 2\pi)]}{\sin[(\alpha + 2\pi)/2]} = (-)^{2\ell}\chi^\ell(\alpha) \qquad (5.46a)$$

For integer ℓ we find the expected result, $\chi^\ell(\alpha + 2\pi) = \chi^\ell(\alpha)$. For half-odd integer ℓ (such as occurs for spinning electrons), this rotation gives $\chi^\ell(\alpha + 2\pi) = -\chi^\ell(\alpha)$. However, for this case,

$$\chi^\ell(\alpha \pm 4\pi) = \chi^\ell(\alpha) \qquad (5.46b)$$

Thus, in this event, we may label rotation through 4π the identity, E, and rotation through 2π, the symmetry operation R, so that $R^2 = E$. We then expand any ordinary rotation group by taking products of R with all existing rotations. The new group has twice as many operations and more classes (but not twice as many) than the starting group, and is called a *double space group*, which conventionally is labeled with a star.

Consider, for example, the D_4 point group which includes eight symmetry operations that separate into five classes: $E, (C_4, C_4^3), C_2, 2C_2', 2C_2''$. The related double group, D_4^*, contains sixteen elements which partition into seven classes:

$$E, R, (C_4^3, RC_4^3), (C_4, RC_4), (C_2, RC_2), (2C_2', 2RC_2'), (2C_2'', 2RC_2'')$$

As there are seven classes in this double group, it must likewise have seven irreps. The dimensions of these irreps satisfy the summational rule (3.27)

$$\ell_1^2 + \ell_2^2 + \cdots + \ell_7^2 = 16$$

The only set of positive integers satisfying this relation is $(1, 1, 1, 1, 2, 2, 2)$. It follows that the double group D_4^* has four one-dimensional and three two-dimensional irreps. As with the direct space group, double space groups likewise contain the translation group as an invariant subgroup [cf., (5.32)].

Spin-Orbit Interaction

The spin-orbit interaction in an atom refers to the interaction of the spin of a valence electron and the magnetic field due to its passage through the Coulomb field of the inner electron shells and the interior nucleus. The related component Hamiltonian is given by

$$H_{SO} = \frac{\hbar}{2m^2c^2}\mathbf{S} \cdot [\nabla(\mathbf{r}) \times \mathbf{p}] \tag{5.47a}$$

where $\mathbf{S}$ is electron spin, $V(\mathbf{r})$ is the Coulomb field of inner electron shells and the nucleus and $\mathbf{p}$ is electron momentum. Due to the translational invariance of $V(\mathbf{r})$ and the fact that $\mathbf{S}$ is independent of translations, H_{SO} is likewise invariant to translations and we write

$$[H_{SO}, \mathbf{T}] = 0 \tag{5.47b}$$

Symmorphic Crystals with Inversion

We wish to discover the effects of spin-orbit coupling on symmorphic crystals with inversion symmetry. Spin transforms as the $\mathbf{D}^{1/2}$ irrep (Section 4.4). Let Γ_i represent an irrep of the group for a given energy band that does not include spin-orbit coupling. The corresponding double-group irrep for this band that includes spin-orbit interaction is given by the direct product, $\mathbf{D}^{1/2} \otimes \Gamma_i$. For example, in simple cubic or P, I, and F cubic lattices, (Table 5.2) the group of $\mathbf{k}$ at the Γ point in the Brillouin zone is

O_h. The group of $\mathbf{k}$ at the R point (Fig. 5.9) in the Brillouin zone is likewise O_h. With reference to the character tables (Appendix A) we see that
there are 10 irreps for the O_h group, five labeled 'g' and five labeled 'u.'
The (u, g) basis functions are (even, odd) with respect to inversion. The
g irreps are relabeled Γ_1^+ to Γ_5^+ and the u irreps are relabeled Γ_1^- to Γ_5^-
(Bethe notation). In this context, $\mathbf{D}^{1/2}$ is relabeled, Γ_6^+, and one finds the
following list of direct products for the corresponding double-group irreps:

$$\Gamma_6^+ \otimes \Gamma_1^\pm = \Gamma_6^\pm, \quad \Gamma_6^+ \otimes \Gamma_4^\pm = \Gamma_6^\pm + \Gamma_8^\pm$$

$$\Gamma_6^+ \otimes \Gamma_2^\pm = \Gamma_7^\pm, \quad \Gamma_6^+ \otimes \Gamma_5^\pm = \Gamma_7^\pm + \Gamma_8^\pm \qquad (5.47c)$$

$$\Gamma_6^+ \otimes \Gamma_3^\pm = \Gamma_8^\pm$$

Appendix B lists the irreps of the O_h and D_{4h} groups in three notations:
Chemical, Bethe, and BSW, the latter being due to Bouchaert, Smo-
luskowski and Wigner. As an example of the application of the products
(5.47c) consider the fcc crystals, germanium and silicon whose symmetries are given by the cubic O_h group. (Diamond also has an fcc lattice.
However, as discussed in Section 5.5, the corresponding space group is non-
symmorphic.) The state at the Γ point ($k = 0$) at the top of the valence
band transforms as the Γ_5^+ irrep of the O_h group. The effect of spin-orbit
coupling on this state is given by the direct product of Γ_5^+ and Γ_6^+ which,
with (5.47c), is seen to split the corresponding energy level into twofold
degenerate Γ_7^+ and fourfold degenerate Γ_8^+ levels (see Appendix B). In the
starting direct product, Γ_6^+ is 2-fold degenerate and Γ_5^+ is 3-fold degenerate so that the direct product of these two irreps has sixfold degeneracy.
These six degenerate states partition into two states of the Γ_7^+ irrep and
four states of the Γ_8^+ irrep.

Solid-State Description

In the solid phase, single-crystal silicon and germanium have the diamond
structure (Fig. 5.11) in which each atom is surrounded by four others located at the vertices of a tetrahedron. The lattice is fcc and the basis
is composed of two atoms with one located at $(0, 0, 0)$ and the other at
$(1/4, 1/4, 1/4)$. [The corresponding reciprocal lattice space is bcc]. The
electronic structures of these atoms are (Ne) $3s^2 3p^2$ for silicon and (Ar)
$3d^{10} 4s^2 4p^2$ for germanium. In the solid state, for each material, the four
outer electrons per atom produce electron bonds with four other atoms in
which the s and p orbitals form hybrid wavefunctions that give rise to four
equivalent bonds. The s state ($\ell = 0$) is non-degenerate. The p state ($\ell = 1$)
is threefold degenerate (the dimension of the Γ_5^+ irrep). Under spin-orbit
coupling, spin $1/2$ adds to the s and p states to give total angular momenta
$j = 1/2, 3/2$. These j values give twofold and fourfold degeneracy, corre-

sponding, respectively, to the dimensions of the Γ_7^+ and Γ_8^+ irreps (Fig. 5.15).

The O_h^* Double Group

With the preceding discussion we are prepared to list the dimensions of the irreps of the O_h^* double group. With Appendix A we see that dimensions of the irreps of O_h are 1, 1, 2, 3, 3. The sum of the squares of these terms gives the order 24 of the group. With reference to (5.47c) we note that direct products of Γ_6^+ with any of the irreps of O_h give the irreps Γ_6^+, Γ_7^+, or Γ_8^+. The dimensions of these irreps are, respectively, 2, 2, and 4, whose squares likewise add to 24. It follows that the dimension of O_h^* is 48.

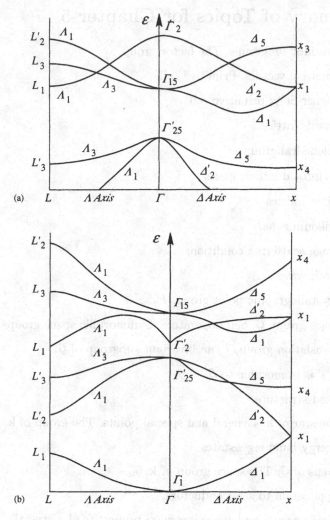

Figure 5.15. Calculated band structure (in the absence of spin-orbit interaction), for: (a) silicon and (b) germanium, in k-space, in the Λ and Δ directions (Fig. 5.9(b)) showing the degeneracy of the Γ'_{25} state (BSW notation) at the Γ point at the top of the valence band. Each curve terminates at an X, of L point, corresponding to Λ and Δ directions, respectively. Branch curves are dileneated with subscripts. [H. Herman, *Phys. Rev.* **95**, 847 (1954).]

Summary of Topics for Chapter 5

1. Invariant subgroups. The factor group.

2. Primitive vectors. Primitive cell.

3. Wigner-Seitz primitive cell.

4. Bravais lattice.

5. Holohedral groups.

6. Reciprocal lattice space.

7. Miller indices.

8. Brillouin zone.

9. Bragg scattering conditions.

10. Bloch waves.

11. Crystallographic point group, P_g.

12. Space group, $\mathcal{G}$, Seitz operators. Symmorphic space groups.

13. Translation group, $\bar{T}$, an invariant subgroup of $\mathcal{G}$.

14. $\mathcal{G}/\bar{T}$ is isomorphic to $\mathcal{G}$.

15. Band structure.

16. The star of $\mathbf{k}$. General and special points. The group of $\mathbf{k}$, $P(\mathbf{k})$.

17. Energy band eigenstates.

18. Irreps of $\mathcal{G}$. The space group of $\mathbf{k}$, $\mathcal{G}_{\mathbf{k}}$.

19. Application to semiconductors.

20. Time-reversal and space-inversion properties of a crystal.

21. Double space groups.

Problems

5.1 Let pS and qS be two left cosets of the subgroup S. Show that either pS and qS are identical or are disjoint.

5.2 Show that the left and right cosets of an invariant subgroup are equal.

5.3 Show that any subgroup, S, of an Abelian group, $\mathcal{G}$, is an invariant subgroup.

Answer

Let $S_i \in S, G_k \in G$. Then,

$$G_k^{-1} S_i G_k = G_k^{-1} G_k S_i = S_i$$

5.4 (a) Write down the C_{4v} group table. (b) How many subgroups does this group have?

Answer (partial)

(a) Pictured as a square matrix, this group table is a symmetric matrix given by

C_{4v}	E	C_4	C_4^2	C_4^3	σ_a	σ_b	σ_c	σ_d
E	E	C_4	C_4^2	C_4^3	σ_a	σ_b	σ_c	σ_d
C_4	C_4	C_4^2	C_4^3	E	σ_b	σ_c	σ_d	σ_a
C_4^2	C_4^2	C_4^3	E	C_4	σ_c	σ_d	σ_a	σ_b
C_4^3	C_4^3	E	C_4	C_4^2	σ_d	σ_a	σ_b	σ_c
σ_a	σ_a	σ_b	σ_c	σ_d	E	C_4	C_4^2	C_4^3
σ_b	σ_b	σ_c	σ_d	σ_a	C_4	E	C_4^3	C_4^2
σ_c	σ_c	σ_d	σ_a	σ_b	C_4^2	C_4^3	E	C_4
σ_d	σ_d	σ_a	σ_b	σ_c	C_4^3	C_4^2	C_4	E

5.5 Show that the purely rotational I subgroup of the I_h group (relevant to the icosahedron and dodecahedron) is not an invariant subgroup.

5.6 Consider the C_{4v} group. (a) What is the smallest invariant subgroup of this group of order $h > 1$? Show that your choice is an invariant subgroup. (b) Construct the factor group, C_{4v}/S, corresponding to this invariant subgroup. (c) What is the order, $h(S)$, of this factor group? (d) Cite two characteristics of a factor-group table. (e) Write down the group table for this factor group. (f) Consider the C_{nv} group (n even). What is the invariant subgroup of minimum order? (g) What is the invariant subgroup of maximum order? (h) In both the preceding cases, what are the orders of the respective factor groups?

Answers (partial)

(a) With reference to the preceding C_{4v} group table, the smallest invariant subgroup is $S = (E, C_4^2)$. For any element $X \in G, X^{-1}S = S$. (b) The corresponding factor group, C_{4v}/S, has components $[S, \{C_4, C_4^3\}, \{\sigma_v, \sigma_v'\}, \{\sigma_d, \sigma_d'\}] \equiv [S, C, \Sigma_v, \Sigma_d]$. (c) The order of the factor group is $4 = 8/2$, in accord with (5.7b). (d) Viewed as a square matrix, this matrix is symmetric with E along its diagonal. (Every element is its own inverse and $S = E$.)

5.7 (a) Consider the groups in Table 5.4. List which groups are subgroups of others in this list. (b) To which group or groups is the factor group $\mathcal{G}/\bar{T}$ isomorphic?

5.8 Draw a square array of lattice points in two dimensions. Insert on this array three different primitive cells and state how many lattice points each includes.
Answer
If the cell has no lattice points on its surface, then it includes one lattice point. If not, it includes more than one lattice point.

5.9 What are the respective Miller indices of the six faces of a cube of edgelength a, with edges aligned with the three Cartesian axes and situated in the first octant? *Note*: The Miller index of an intercept at infinity is zero.

5.10 Employing Bragg's interference equations, derive the interference equation (5.20c).
Answer
Consider the figure shown below:

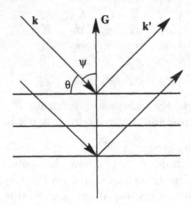

where $\mathbf{G}$ is shown normal to the corresponding (hkl) scattering planes of the problem and scattering is elastic $(k = k')$. From (5.20b),

$$2kG \cos \varphi = G^2$$
$$2k \sin \theta = |G|$$

With the formula for distance between parallel (hkl) scattering planes (5.16b), and setting $k = 2\pi/\lambda$, we obtain

$$2(2\pi/\lambda) \sin \theta = \frac{2\pi}{d(hkl)}$$

To delineate parallel planes Miller indices are written in unfactored form $(1/n)(h, k, l)$ where n is an integer. In this case, $d(h, k, l) \rightarrow d/n$. Substituting this expression into the preceding equation returns (5.20c).

5.11 Establish (5.32b).

Answer

$$\{R|\mathbf{T}\}\{E|\mathbf{T}\}\{R|\mathbf{T}\}^{-1}\mathbf{r} = \{R|\mathbf{T}\}\{E|\mathbf{T}\}\{R^{-1}| - R^{-1}\mathbf{T}\}\mathbf{r}$$
$$= \{R|\mathbf{T}\}\{E|\mathbf{T}\}[R^{-1}\mathbf{r} - R^{-1}\mathbf{T}]\mathbf{r} = \{R|\mathbf{T}\}[R^{-1}\mathbf{r} - R^{-1}\mathbf{T} + \mathbf{T}]\mathbf{r}$$
$$= R[R^{-1}\mathbf{r} - R^{-1}\mathbf{T} + \mathbf{T}] + \mathbf{T} = \mathbf{r} - \mathbf{T} + R\mathbf{T} + \mathbf{T} = \mathbf{r} + R\mathbf{T}$$
$$= \mathbf{r} + \mathbf{T}' = \{E|\mathbf{T}'\}\mathbf{r}$$

where $\mathbf{T}' = R\mathbf{T}$ is a primitive translation vector. That is, elements of the translation group are maintained under the given similarity transformation of elements of $\mathcal{G}$, so that $\bar{T}$ is an invariant subgroup of $\mathcal{G}$.

5.12 What is the relation between the space group $\mathcal{G}$, the space group of $\mathbf{k}, \mathcal{G}_\mathbf{k}$, and the group of $\mathbf{k}, P(\mathbf{k})$?

Answer

$$P(\mathbf{k}) \subset \mathcal{G}_\mathbf{k} \subset \mathcal{G}$$

5.13 For a general $\mathbf{k}$, the group of $\mathbf{k}$ of a certain crystal has one element, the identity. What can be said about the space group for this crystal?

5.14 For symmorphic groups, what can be said about the basis functions of the irreps of the space group of $\mathbf{k}$, $\mathcal{G}_\mathbf{k}$, and basis functions of the irreps of the group of $\mathbf{k}, p(\mathbf{k})$?

5.15 Show that the alternating group (Section 4.5) of order $n \geq 4$ has no invariant subgroups except itself and the identity.

5.16 Show that the alternating group of degree $(n!/2)$ is an invariant subgroup of the symmetric group $\mathcal{S}_n$.

5.17 With reference to (5.40b), write down an explicit representation of the matrix $\Gamma^{\mathcal{G}}_{ij}(\{R|\mathbf{T}\})$.

Answer

$$\Gamma^{\mathcal{G}}_{ij} = \langle \varphi_{n_j k_j} | \{R|\mathbf{T}\} \varphi_{n_i k_i} \rangle$$

where the degenerate set, $\{\varphi_{n_j k_j}\}$, is assumed to be orthonormalized.

5.18 A group is said to be *simple* if it has no invariant subgroups. Is the C_{nv} group simple?

5.19 (a) Show that the reciprocal of a Bravais lattice is a Bravais lattice. *Hint:* Show that the reciprocal of a reciprocal lattice is the direct lattice. (b) Show that a direct lattice and its reciprocal are described by the same point groups. (c) Show that a direct lattice and its reciprocal are described by the translation group.

Answer (partial)

(b) The point group of a lattice is composed of rotations, improper rotations, reflections and inversions. In the following we introduce the acronyms, RLS and RLV for reciprocal lattice space and a reciprocal lattice vector, $\mathbf{G}$, respectively.

Consider a finite regular n-gon prism with C_n ($n \geq 2$) symmetries. For n even, there is a normal vector to each of the $2 + n$ faces of the prism, where C_2 refers to rotation of the principal axis of the prism through π about its center. For n odd, there are n rotations of the prism about the principal axis. In either case, in the C_n operation, all normals are mapped onto each other. In the C_2 operation, two normals are exchanged and the remaining normals are invariant. Each such normal is an RLV in RLS. So in all, these operations in direct space map RLS onto itself.

Symmetry of a polygon plane of bisecting symmetry of the prism and C_n symmetry are described by the improper rotation operator S_n (Table 1.1). Consider the $\mathbf{G}$ vector normal to such a plane. The S_n operator maps this $\mathbf{G}$ onto $-\mathbf{G}$ which is also an element of an RLS. We conclude that improper rotations map reciprocal lattice space onto itself. As $S_2 = i$, the inversion operator, we conclude that inversions likewise map reciprocal lattice space onto itself. One may conclude that the point group of the direct lattice maps reciprocal lattice space onto itself.

(c) For a translation $\mathbf{T}$ [given by (5.8b)], with (5.15e) we write

$$\mathbf{b}_1' = \frac{(\mathbf{a}_2 + \mathbf{T}) \times (\mathbf{a}_3 + \mathbf{T})}{V_0} \equiv \mathbf{b}_1 + \Delta$$

Solving for Δ one finds that

$$\mathbf{b}_1' = N_1 \mathbf{b}_1 + N_2 \mathbf{b}_2 + N_3 \mathbf{b}_3$$

Where $\{N_k\}$ are integers. Thus, to any primitive translation vector in direct space, there corresponds an RLV in RLS and vice versa. It follows that primitive translations of a direct lattice and RLVs in reciprocal lattice space are described by the translation group.

5.20 (a) How many point groups are there? (b) How many crystallographic point groups are there? (c) Offer an explanation for the difference. (d) If Σ represents the set of point groups and P_g the set of crystallographic point groups, write down a set-theoretic relation between these two sets.

Answers

(a) There are a countable infinity of point groups. (b) There are 32 crystallographic point groups. (c) There are only 14 Bravais lattices. (d) $P_g \subset \Sigma$.

5.21 Derive the group table for the nonsymmorphic group D_3^4 listed in (5.30).

5.22 If G is any group, then show that E and G are invariant subgroups of G.

Answer

$$G_k^{-1} G_j G_k = G_q \in G$$
$$G_k^{-1} E G_k = E$$

6

Atoms in Crystals and Correlation Diagrams

6.1 Central-Field Approximation

Consider an atom with Z protons in its nucleus and with Z outer electrons (the *atomic number* of the atom is Z). In the central-field approximation, atomic electrons are assumed to be independent of one another. The Hamiltonian of the ith atomic electron is given by

$$H_i(\mathbf{r}_i) = \frac{p_i^2}{2m} - \frac{Ze^2}{\mathbf{r}_i} \tag{6.1a}$$

which has the eigenstates

$$\Psi_{n\ell m}(\mathbf{r}_i) = R_{n\ell}(\mathbf{r}_i)Y_{\ell m}(\theta_i, \phi_i)\xi_i \tag{6.1b}$$

The spherical harmonic functions, $Y_{\ell m}(\theta, \phi)$, were encountered previously (cf., (4.15); (4.36b)). It was observed that at any ℓ, these functions form a $(2\ell+1)$ dimensional basis of the irreps of the rotation $O(3)^+$ group. The radial component wavefunctions, $R_{n\ell}(\mathbf{r})$, are exponentially damped Laguerre polynomials[1] and ξ_i represents the spin function of the ith electron. In the central-field approximation the atomic Hamiltonian is given by

$$H(\mathbf{r}_1, \mathbf{r}_2, \dots, \mathbf{r}_Z) = H_1(\mathbf{r}_1) + H_2(\mathbf{r}_2) + \dots + H_Z(\mathbf{r}_Z) \tag{6.1c}$$

---

[1]See, for example, R.L. Liboff, *Introductory Quantum Mechanics*, 4th ed. Addison-Wesley, San Francisco, CA (2002), Table 10.3.

Wavefunctions are given by the product form

$$\Psi_{n_1 n_2 \ldots n_Z}(r_i, r_2, \ldots, r_Z) = \Psi_{n_1}(r_1)\Psi_{n_2}(r_2)\ldots\Psi_{n_Z}(r_Z) \qquad (6.1d)$$

and eigenenergies by the sum

$$E = E_1 + E_2 + \ldots + E_Z \qquad (6.1e)$$

In (6.1d) the index n_i denotes the quantum number set (n_i, ℓ_i, m_i, m_s), where m_s is the projection spin eigenvalue, $\pm 1/2$. This set of quantum values together with the Pauli principle imply a shell structure of atoms.

Electron and Atomic State Notation

Angular momentum ℓ values of individual atomic electrons carry the following notation:

ℓ-value	0	1	2	3	4	5
Letter	s	p	d	f	g	h

Thus, for example, the ground-state electronic configuration of the helium atom is written $1s^2$, indicating that there are two s electrons with principal quantum number 1. The ground-state electronic configuration of the boron atom is written $1s^2 2s^2 2p^1$, indicating that there are two s electrons with principal quantum number 1, and two s electrons with principal quantum number 2, and one p electron with principal quantum number 2.

In the *Russell-Saunders* coupling scheme, it is assumed that spins of individual electrons couple to one another to give a net **S** value and orbital angular momentum of individual electrons couple to one another to give a net **L** value. These two vectors couple to give a total angular momentum **J** value, see (4.35),

$$\mathbf{J} = \mathbf{L} + \mathbf{S} \qquad (6.2a)$$

Angular momentum eigenvalues have the form

$$J^2 = \hbar^2 j(j+1); \quad L^2 = \hbar^2 \ell(\ell+1); \quad S^2 = \hbar^2 s(s+s) \qquad (6.2b)$$

If an atom is in a state with given ℓ and s, then possible j values are given by

$$j = |\ell - s| \ldots |\ell + s| \qquad (6.2c)$$

in unit steps. For $\ell < s$, there are $2\ell + l$ values of j. For $s < \ell$, there are $2s + l$ values of j. This total number of j values is called the *multiplicity* of the state. For example, in a two-electron atom such as helium, possible s values are $s = 0, 1$. For any given ℓ value of the atom, these s values give rise to a singlet series, $j = \ell$, and a triplet series, $j = \ell - 1, \ell, \ell + 1$.

Orbital angular momentum states of an atom likewise carry a letter equivalence parallel to that given above for angular momenta of individual electrons. This scheme is given by the following list:

L-value	0	1	2	3	4	5
L-letter	S	P	D	F	G	H

Atomic states are designated by the symbol $^{(2s+1)}L_j$. The term $2s+1$ indicates the multiplicity of the state, L the letter equivalent of the orbital angular momentum quantum number and j the total angular momentum quantum number of the state. The singlet and triplet series described above, say for the $\ell = 2$ state, are written $^1D_2; {}^3D_{1,2,3}$, respectively.

6.2   Atoms in Crystal Fields

In this section we consider the influence on degeneracies of an atom placed in a crystal field. As noted previously in Section 4.1, degeneracy of the quantum state is related to the symmetries of the system. In the central-field approximation, each atomic electron is in the presence of a spherically symmetric Coulomb potential of the nucleus. Wavefunctions are given by (6.1b), which include the spherical harmonic function $Y_{\ell m}(\theta_i, \phi_i)$. Each such state is $(2\ell + l)$-fold degenerate, corresponding to the azimuthal quantum number m values $(-\ell, \dots, \ell)$ [see (4.15c)] at a given value of ℓ. This degeneracy corresponds to the spherical symmetry of the system. The single value of energy of a degenerate state will split into a number of energy levels, if the symmetry of the system is reduced.

Thus, when an atom is placed in a crystal, the crystal field reduces the symmetry of the atom and degeneracies are partially removed. We wish specifically to determine the manner in which energy levels of, say, a d electron ($\ell = 2$) split in an octahedral environment. The set of five d wavefunctions may be used as a basis for a representation of the point group of the crystal environment to discover the manner in which the five levels split by the crystal field. As the octahedral group, $O_h = O \otimes C_i$; information of the d wavefunctions may be obtained from the pure rotational O subgroup, as any of the d wavefunctions are even under inversion. We wish to find a representation for which the d wavefunctions form a basis of a representation of the symmetry group of the crystal environment. We seek matrices which describe the effects of the symmetry operations of the group on the d wavefunctions. The characters of these matrices are the characters of the representation being sought.

A general expression for the character of an element of the O group corresponding to rotation through γ at a given ℓ value is given by [recall (4.19)],

$$\chi^\ell(\gamma) = \frac{\sin[(\ell + \tfrac{1}{2})\gamma]}{\sin(\gamma/2)} \tag{6.3}$$

As noted in Section 5.3 [(5.12a), et seq.], only the rotations C_2, C_3, C_4, C_6 are relevant to the crystallographic groups. A list of $\chi_\ell(C_n)$ values, where $n = 2, 3, 4, 6$, is given in Table 6.1.

Table 6.1

Values of χ_ℓ (C_n) Relevant to Crystallographic Groups with γ-Values on the Horizontal and ℓ-Values on the Vertical

	0	1	2	3
π	1	-1	1	1
$2\pi/3$	1	0	-1	1
$\pi/2$	1	1	-1	-1
$\pi/3$	1	2	1	-1

Consider the character table of the O group (Fig. 6.1; see also Appendix A). With $\chi(d) = 5$, one notes that this character as well as the characters of $\ell = 2$ rotations are maintained by the irrep

$$\Gamma_d = E_g + T_{2g} \tag{6.4}$$

The five-fold degeneracy of the d electron state is removed when the atom is placed in a crystal with O_h symmetry. With (6.4) we note that degenerate quantum states are split into the triply degenerate states of the T_{2g} irrep and the doubly degenerate states of the E_g irrep.

O_h	E	$8C_3$	$6C_4$	$3C_2$	$6C_2$	i	$8S_6$	$6S_4$	$3\sigma_h$	$6\sigma_d$
A_{1g}	1	1	1	1	1	1	1	1	1	1
A_{2g}	1	1	-1	1	-1	1	1	-1	1	-1
E_g	2	-1	0	2	0	2	-1	0	2	0
T_{1g}	3	0	1	-1	-1	3	0	1	-1	-1
T_{2g}	3	0	-1	-1	1	3	0	-1	-1	1
A_{1u}	1	1	1	1	1	-1	-1	-1	-1	-1
A_{2u}	1	1	-1	1	-1	-1	-1	1	-1	1
E_u	2	-1	0	2	0	-2	1	0	-2	0
T_{1u}	3	0	1	-1	-1	-3	0	-1	1	1
T_{2u}	3	0	-1	-1	1	-3	0	1	1	-1

Figure 6.1. The O_h group with the pure rotational O group partitioned and corresponding characters of $E_g + T_{2g}$ corresponding to $\chi(d)$ and $\chi_d(C_n)$ values.

More generally, splitting of degenerate 1-electron levels, corresponding to respective orbital quantum number states $0 < \ell < 6$ in an octahedral environment, are shown in Table 6.2.

Table 6.2

Splitting of Single Electron Levels in a Crystal with Octahedral Symmetry

Letter Description of State	ℓ	$\chi(E)$	$\chi(C_2)$	$\chi(C_3)$	$\chi(C_4)$	Irrep Components
s	0	1	1	1	1	A_{1g}
p	1	3	-1	0	1	T_{1u}
d	2	5	1	-1	-1	$E_g + T_{2g}$
f	3	7	-1	1	-1	$A_{2u} + T_{1u} + T_{2u}$
g	4	9	1	0	1	$A_{1g} + E_g + T_{1g} + T_{2g}$
h	5	11	-1	-1	1	$E_u + 2T_{1u} + T_{2u}$
i	6	13	1	1	-1	$A_{1g} + A_{2g} + E_g + T_{1g} + 2T_{2g}$

(From: *Chemical Applications of Group Theory*, 3rd ed., F.A. Cotton, Copyright (1990, F.A. Cotton) Reprinted by permission of John Wiley & Sons, Inc.)

The process of determining the manner in which the degeneracy of an ℓ electron is reduced by a crystal environment has three components: (a) Note the degeneracy of the given ℓ electron state. This determines $\chi(E)$ of the irrep components of the sought representation. (b) Determine the $\chi_\ell(C_n)$ values relevant to the given ℓ value. (c) Consult the rotational component of the character table of the crystallographic group specific to the crystal field in which the atom is placed. Irrep components of the sought representation are combinations of irreps of the rotational component of the crystal group which maintain $\chi(E)$ and $\chi_\ell(C_n)$ values.

Russell-Saunders (or, 'LS') Coupling

In Russell-Saunders coupling, orbital angular momenta of individual electrons in an atom couple to give a net ℓ value, as do spin angular momenta to give a net s value. As described in Section 6.1, these values then combine to give a net j value. The quantum states $^1D_2$; $^3D_{1,2,3}$ developed in Section 6.1 are examples of LS coupling. In this representation, the operators $\{J^2, J_z, L^2, S^2\}$ comprise a complete set of commuting operators. In an alternative representation in the LS scheme, $\{L^2, L_z, S^2, S_z\}$ comprise a complete set of commuting operators. For example, consider an atom with three d electrons ($\ell = 2$). The possible net ℓ values follow from first adding two ℓ values and then adding the third. The first two give $\ell = 0, 1, 2, 3, 4$. Adding the third gives the additional value, $\ell = 5$. For crystal systems which do not act directly on electron spins of an impurity atom, the preceding representation indicates that degeneracy of the atom is the same as that of a single electron in a given ℓ state, that is, $2\ell + 1$. In the present example of three d electrons, in a state of definite ℓ, say, $\ell = 3$ (an F state) the degeneracy of the state is 7, due to m projections of orbital angular momentum.

Energy Level Diagrams

It is conventional in describing symmetries of atomic electron states to use small letters (in analogy with the case for angular momentum). These correspond to irreps of the crystal environment, A, E, T, namely, a, e, t. An example of this notation is given in Table 6.3.

Table 6.3

Splitting of Electron Levels in Crystal Fields of O_h and T_d Symmetry, Respectively

Electron Level	Crystal Symmetry O_h	T_d
s	a_{1g}	a_1
p	t_{1u}	t_2
d	$e_g + t_{2g}$	$e + t_2$
f	$a_{2u} + t_{1u} + t_{2u}$	$a_2 + t_1 + t_2$
g	$a_{1g} + e_g + t_{1g} + t_{2g}$	$a_1 + e + t_1 + t_2$
h	$e_u + 2t_{1u} + t_{2u}$	$e + t_1 + 2t_2$
i	$a_{1g} + a_{2g} + e_g + t_{1g} + 2t_{2g}$	$a_1 + a_2 + e + t_1 + 2t_2$

(From: *Chemical Applications of Group Theory*, 3rd ed., F.A. Cotton, Copyright (1990, F.A. Cotton) Reprinted by permission of John Wiley & Sons, Inc.)

Note that entries in this table have g and u subscripts. As previously described, these correspond to even and odd quantum states, respectively. Subscripts corresponding to the case that the point group of the surrounding environment has no center of symmetry do not come into play. All states whose orbital quantum number, ℓ, is even are even under inversion $(s, d, g, \ldots)$ and carry the g subscript. States whose orbital quantum number is odd $(p, f, h, \ldots)$ are antisymmetric under inversion and carry the u subscript.

It has been noted that free atom levels with $\ell > 1$ are split by crystal environments of O_h and T_d symmetry, respectively, into states described by lower symmetries of two or more states labeled according to the point group describing their transformation properties. Let Δ_0 represent the spread in

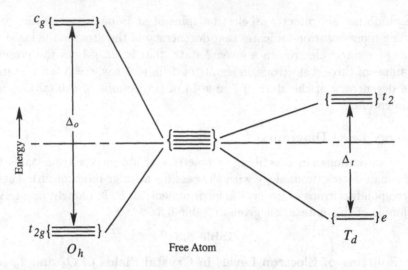

Figure 6.2. Spread in the e and t_2 levels resulting from splitting of the fivefold degenerate d-state by octahedral and tetrahedral environments, relative to the free atom state. (From: *Chemical Applications of Group Theory*, 3rd ed., F.A. Cotton, Copyright (1990, F.A. Cotton). Reprinted by permission of John Wiley & Sons, Inc.)

energy between the e_g and t_{2g} sets of levels of an atom in the d state placed in a crystal in octahedral symmetry. Likewise Δ_t represents the spread in energy between the e and t_2 sets of levels of an atom in the d state placed in a crystal in tetrahedral symmetry. It may be shown[2] that $\Delta_0 > \Delta_t$. This situation is depicted in Fig. 6.2.

6.3   Correlation Diagrams

Consider the change in energy levels of an atom from its free state to the state of infinite crystal coupling in octahedral symmetry. Specifically, let us examine an atom with two d electrons with orbital angular momentum quantum numbers $\ell = 0, 1, 2, 3, 4$ corresponding to the atomic states S, P, D, F, G. When electrons interact with a moderate crystal interaction, V_C, wavefunctions transform as the product representations relevant to the two d electrons

$$e \otimes e = A_1 + A_2 + E, \ \ e \otimes t_2 = T_1 + T_2$$

$$t_1 \otimes t_2 = A_1 + E + T_1 + T_2 \tag{6.5}$$

---

[2]See, for instance, D.M. Bishop, *Group Theory and Chemistry* (see Bibliography).

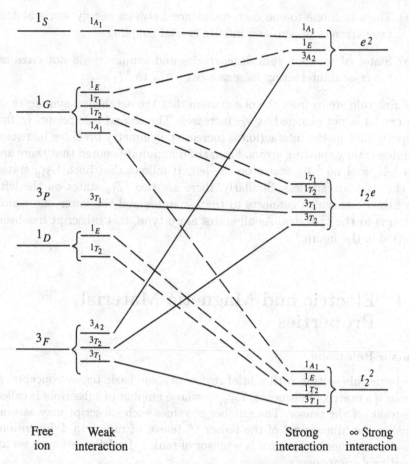

Figure 6.3. Correlation diagram for a d^2 ion in an octahedral environment. As all states are g type, this subscript is omitted. (From: *Chemical Applications of Group Theory*, 3rd ed., F.A. Cotton, Copyright (1990, F.A. Cotton). Reprinted by permission of John Wiley & Sons, Inc.)

All states are g type and this subscript is omitted. As the crystal interaction, V_C, grows large, we have seen that electron wavefunctions transform as $e + t_2$ (Table 6.3) and three electron configurations are possible,

$$t_2^2, \; t_2 e, e^2$$

In this notation, t_2^2 denotes electrons being in t_2 orbitals, etc. These transformations are depicted in the correlation diagram shown in Fig. 6.3. In going from left to right in this diagram there is an increase in crystal interaction and related symmetry breaking. On the extreme left of the diagram are equivalent orbital letters corresponding to the five ℓ values.

Two rules are evident from this diagram:

(a) There is a one-to-one correspondence between energy levels at the two extremes of zero and infinite crystal coupling.

(b) States of common spin degeneracies and symmetry do not cross as the crystal interaction increases (e.g., $^3T_1$ to $^3T_1$).

The first rule stems from the observation that the octahedral symmetry of the crystal is not changed as V_C increases. The second rule relates to the property that as the interaction is increased, symmetry breaking increases and degeneracy splitting grows. From the diagram it is noted that there are two $^1A_{1g}$ and no $^3A_{1g}$ states on the left. It follows that both $^1A_{1g}$ states on the right are singlets. Similarly, there are two $^3T_{1g}$ states on the left, the higher one which connects to the t_2e states and the lower one which connects to the t_2^2 states. As all states are g type, this subscript has been omitted in the figure.[3]

6.4   Electric and Magnetic Material Properties

Tensor Relations

We begin this section with a brief review of some basic tensor concepts. A tensor is a matrix of elements, $T_{mnq...}$, whose number of subscripts is called the 'rank' of the tensor. The number of values each subscript may assume is called the 'dimension' of the tensor. A tensor of rank r in d dimensions has d^r components. A vector is a tensor of rank 1. In d dimensions a vector has $d^1 = d$ components.

Tensors emerge in coordinate transformations. Under rotation of coordinates, new coordinates x_i' are related to original coordinates x_i though the relation (3.7),

$$x_i' = \sum_{j=1}^{3} \Gamma_{ij} x_i \qquad (6.6a)$$

The matrix Γ_{ij} is a second-rank tensor in three dimensions. It has the orthogonality property

$$\sum_{j=1}^{3} \Gamma_{ij} \Gamma_{kj} = \delta_{ik} \qquad (6.6b)$$

---

[3]For further discussion see, A.F. Cotton, *Chemical Applications of Group Theory*, 3rd ed., *ibid* (see Bibliography); D.M. Bishop, *Group Theory and Chemistry* (see Bibliography); B.N. Figgis, *Introduction to Ligand Fields* (see Bibliography).

where δ_{ik} is the Kronecker delta symbol, $\delta_{ik} = 1, i = k, \delta_{ik} = 0, i \neq j$. Components of the δ_{ik} matrix have 1's on the diagonal elements and 0's on the off-diagonal elements.

In the 'Einstein convention,' repeated indices in a product are summed. Thus, for example, in this notation, (6.6b) is written

$$\Gamma_{ij}\Gamma^{\dagger}_{jk} = \delta_{ik} \tag{6.6c}$$

where $\Gamma^{\dagger}$ is written for the Hermitian adjoint of Γ. For real $\Gamma, \Gamma^{\dagger} = \tilde{\Gamma}$, the transpose of Γ.

To find the manner in which a second-rank tensor transforms, we consider the *diad*, $x_i x_j$ which transforms as

$$x'_i x'_j = \Gamma_{ik} x_k \Gamma_{jm} x_m = \Gamma_{ik} \Gamma_{jm} x_k x_m \tag{6.7a}$$

This relation indicates the manner in which second-rank tensors transform, namely,

$$A'_{ij} = \Gamma_{ik}\Gamma_{jm}A_{km} \tag{6.7b}$$

More generally, an rth-rank tensor transforms as

$$A'_{ijk...} = [\Gamma_{im}\Gamma_{jn}\Gamma_{ko}\cdots]A_{mno...} \tag{6.7c}$$

Two vectors are related through a second-rank tensor. For example, in material physics, electric polarization $\mathbf{P}$ and electric field $\mathbf{E}$ are related through the *polarizability tensor* α_{ij} as follows

$$P_i = \alpha_{ij} E_j \tag{6.7d}$$

Constitutive Relations

In the following, $\mathbf{E}$ and $\mathbf{D}$ are electric fields and $\mathbf{B}$ and $\mathbf{H}$ are magnetic fields. These fields are related through Maxwell's equations.[4] To describe the influence of material properties on these fields, the electric polarization field, $\mathbf{P}$, and magnetic polarization field, $\mathbf{M}$, are introduced and are defined by the following constitutive relations (in cgs units)

$$\mathbf{D} = \mathbf{E} + 4\pi\mathbf{P} \equiv \mathbf{E}(1 + 4\pi\chi_e) \equiv \epsilon\mathbf{E} \tag{6.8a}$$

$$\mathbf{P} = \chi_e\mathbf{E} \tag{6.8b}$$

$$\mathbf{B} = \mathbf{H} + 4\pi\mathbf{M} = \mathbf{H}(1 + 4\pi\chi_m) \equiv \mu\mathbf{H} \tag{6.8c}$$

$$\mathbf{M} = \chi_m\mathbf{H} \tag{6.8d}$$

The parameters ϵ, μ are electric and magnetic permeabilities, respectively, and are related to the corresponding susceptabilities, χ_e, χ_m, as

---

[4]For further discussion, see J.D. Jackson, *Classical Electrodynamics*, 3rd ed., Wiley, New York (1999).

implied. These equations apply to bulk matter. As with polarizability, susceptabilities are likewise second-rank tensors.

Neumann's Principle

Neumann's principle is important to the study of symmetries in material media. It states: any physical property of a crystal possesses the symmetry of the point group of the crystal. Thus, if the tensor T represents a property of the material, the value of each tensor component $T_{ijk...}$ is invariant under the group of symmetries of the material.

Electric Group Properties

Crystals whose primitive cells have a nonvanishing electric dipole moment, $\mathbf{p}_0$, are called *pyroelectric*. For such a crystal, if the total number of primitive cells is N, and the volume occupied by the crystal is V, then the corresponding polarization vector has the value $\mathbf{P} = N\mathbf{p}_0/V$. For such a crystal the only rotation axis is in the direction of $\mathbf{P}$. Furthermore, there can be no mirror planes perpendicular to that axis (the vector $\mathbf{P}$ reverses sign under such reflection). Point groups satisfying these constraints are the C_{nv} groups for $n = 2, 3, 4, 6$, and the single-element group C_1 (Chapter 1). These groups are consistent with Neumann's principle as in any C_{nv} group, z transforms as the totally symmetric A_1 irrep, so that crystals described by any of these point groups can have polarization only along the z axis.

Polarizability

The polarizability tensor α_{ij} connects two vectors ($\mathbf{E}$ and $\mathbf{P}$) and is itself a second-rank tensor and has nine components. Energy considerations imply that $\alpha_{ij} = \alpha_{ji}$, that is, α_{ij} is a symmetric tensor. Additional constraints on these elements stem from the property that components of a second-rank tensor transform as the product of orthogonal Cartesian coordinates two at a time. Furthermore, Neumann's principle indicates that any symmetry operation which gives $x_i x_j = -x_i' x_j'$ has the corresponding $\alpha_{ij} = 0$, as the principle indicates that any component must transform into $+1$ times itself.

We wish to find the representation of the polarizability tensor for a crystal with C_{2h} symmetry. Recall that this group has elements $C_{2h} = (E, C_2, i, \sigma_h)$. The character table for this group indicates that the six element of α_{ij} transform as $x^2, y^2, z^2, xy, xz, yz$. The C_2 operation transforms x into $-x, y$ into $-y$ and z into $+z$. It follows that under this transformation $xz \rightarrow -xz$ and $yz \rightarrow -yz$, which with Neumann's principle indicates that $\alpha_{13} = \alpha_{23} = 0$. As the α_{ij} tensor is symmetric it follows that $\alpha_{31} = \alpha_{32} = 0$. Remaining components are left invariant. Similarly, the i

and σ_h transformations leave products invariant. This leaves the matrix

$$\bar{\alpha} = \begin{pmatrix} \alpha_{11} & \alpha_{12} & 0 \\ \alpha_{12} & \alpha_{22} & 0 \\ 0 & 0 & \alpha_{33} \end{pmatrix} \tag{6.9}$$

where $\bar{\alpha}$ represents the α_{ij} tensor. In similar manner one finds that α_{ij}, for a crystal with cubic symmetry, is diagonal with $\alpha_{11} = \alpha_{22} = \alpha_{33}$.

Piezoelectric Effect

The piezoelectric effect refers to the phenomenon of a strain being developed in a crystal in response to the application of an electric field or, conversely, an electric polarization field being developed in response to a strain field in a crystal. Unlike a pyroelectric crystal, a peizoelectric crystal has no permanent dipole moment. Strain e_{ij} is a symmetric second-rank tensor with at most six independent components. It is a measure of incremental displacements in the crystal in response to applied forces, so that, for example, e_{xy} is the x incremental displacement (shear) of a surface element whose normal is in the y direction. As strain is a second-rank tensor and electric field is a vector, the coefficient relating the two, the 'piezoelectric constant,' d_{ijk}, is a third-rank tensor and we write

$$e_{jk} = d_{ijk}E_i \tag{6.10}$$

There are in general $3^3 = 27$ components of the d_{ijk} tensor. The term d_{xyz}, for example, represents the e_{yz} element of strain developed as a result of an $\mathbf{E}$ field in the x direction. The d_{ijk} tensor is symmetric so that, $d_{ijk} = d_{ikj}$, as the strain tensor is symmetric. Thus, in semiconductors with sphalerite of wurtzite structure (but not rocksalt structures) elastic strain may be accompanied by macroscopic electric fields. A crystal with inversion symmetry is not piezoelectric. This is so because under the i operation, $x_i x_j x_k \rightarrow -x_i' x_j' x_k'$ and with Neumann's principle, all components of d_{ijk} vanish.

As above, the manner in which zero elements of the d_{ijk} constants are determined is obtained from the property that components of a third-rank tensor transform as the product of orthogonal Cartesian coordinates three at a time. Thus one must ascertain the manner in which the terms x_i, x_j, x_k transform under the operation of the point group relevant to the crystal at hand.

Consider, for example, a crystal with D_2 symmetry. This group has elements $(E, C_{2x}, C_{2y}, C_{2z})$. Again, restrictions on tensor components may be inferred with Neumann's principle and the generators of the D_2 group, namely, C_{2z} and C_{2y}. Consider first the element C_{2z}, for which $x^3 \rightarrow -x^3, y^3 \rightarrow -y^3$ so that with Neumann's principle, $d_{111} = d_{222} = 0$. Similarly,

$$yx^2 \rightarrow -yx^2, \; xy^2 \rightarrow -xy^2, yz^2 \rightarrow -yz^2, xz^2 \rightarrow -xz^2 \tag{6.11a}$$

which imply $d_{211} = d_{122} = d_{233} = d_{133} = 0$. Under the C_{2y} operation we obtain the additional relations

$$zy^2 \rightarrow -zy^2, zx^2 \rightarrow -zx^2, z^3 \rightarrow -z^3 \qquad (6.11b)$$

which imply $d_{322} = d_{311} = d_{333} = 0$. In this manner one obtains the nine conditions

$$d_{322} = d_{311} = d_{122} = d_{211} = d_{233} = d_{133} = 0$$
$$d_{111} = d_{222} = d_{333} = 0 \qquad (6.12)$$

Note that $yx^2 \rightarrow -yx^2$ corresponding to $d_{211} = 0$ may be rewritten $xyz \rightarrow -xyx$ which gives $d_{121} = 0$. Due to symmetry to the strain tensor, the first relation in (6.12) implies the relations $d_{322} = d_{232} = 0$. Permuting xyz as described, gives the additional relation $d_{223} = 0$. Thus the top line of (6.12) gives 18 elements which are zero. Together with the bottom line this gives a total of 21 elements. Only the product xyz and its permutations transform as A_1 which gives six non-vanishing d_{ijk} components out of a total of $27 = 3^3$ elements of the peizoelectric component tensor. For example $d_{123} = d_{132}$ corresponds to yz strain $(= zy$ strain$)$ developed in the crystal by an electric field in the x direction.

Magnetic Properties

Magnetic materials divide into the following primary categories: diamagnetic, paramagnetic and ferromagnetic. The first two of these are defined with respect to magnetic susceptibility:

$$\chi_m < 0, \quad \text{diamagnetic} \qquad (6.13a)$$

$$\chi_m > 0, \quad \text{paramagnetic} \qquad (6.13b)$$

A ferromagnetic material has a magnetic moment in the absence of an applied magnetic field which may be associated with aligned spins. An antiferromagnetic material has spins antialigned with no net magnetic moment (Fig. 6.4). With (6.8d), the condition (6.13a) indicates that the magnetization, $\mathbf{M}$, of a diamagnetic material decreases in the presence of an applied magnetic field.

With (6.13b) it follows that the magnetization of a paramagnetic material increases in the presence of an applied magnetic field. Atoms and molecules are diamagnetic.

The sources of a magnetic field are currents and spins of elementary charges. If internal charge currents in a crystal do not average to zero, a magnetic field will emerge. Suppose the crystal has symmetry described by a point group G. Under any element of this point group, the crystal is brought into geometrical coincidence with itself. However, it may be the case that this operation reverses currents and/or spins. It follows that the symmetry operation by itself is not a symmetry operation of the crystal. To

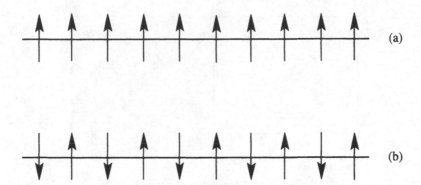

Figure 6.4. One-dimensional representations of (a) a ferromagnetic crystal and (b) an antiferromagnetic crystal. Magnetic moments are represented by arrows.

remedy this situation one introduces an operator $\mathcal{M}$, which when operating on the crystal reverses the sign of all current densities and magnetic moments at all points of the crystal. This operator is termed the *antisymmetry* or *time-reversal operator*. It does not act on the space coordinates of the crystal and is its own inverse, $\mathcal{M}^2 = E$. As $\mathcal{M}$ is basically a two-component operator, one may envision it as changing $+$ to $-$, black to white, reversing the direction of time, or changing spins from up to down. The operation $\mathcal{M}$ commutes with all point-group symmetry operations.

Classification of Magnetic Crystals

There are three classifications of magnetic crystals. The first of these has crystals which are described by members of the 32 (P_g) point groups discussed above with no $\mathcal{M}$ operations. These include C_4 or C_{4h} groups, as all operations in these groups transform 'spin up' positions among themselves. Applying $\mathcal{M}$ for such cases reverses the sign of spins and is not a symmetry operation.

The second classification of crystals are described by any of the 32 point groups $\{\mathcal{A}_i\}$ together with the operations $\{\mathcal{M}\mathcal{A}_i\}$. The group is given by $\{\mathcal{A}_i\} + \{\mathcal{M}\mathcal{A}_i\}$. If h is the order of the point group $\{\mathcal{A}_i\}$, then the extended group has order $2h$. Note that $\mathcal{M}$ is an element of this group. As $\mathcal{M}$ reverses magnetic moments at all points, paramagnetic and diamagnetic crystals are included in this category. These groups are sometimes called 'grey groups,' as they mix 'black and white' groups.

In the third classification of magnetic crystals, $\mathcal{A}_i$ by itself is not a symmetry operation but $\mathcal{M}\mathcal{A}_i$ is included. There are 58 such groups and their properties are described below and they are sometimes referred to as 'black and white' groups.

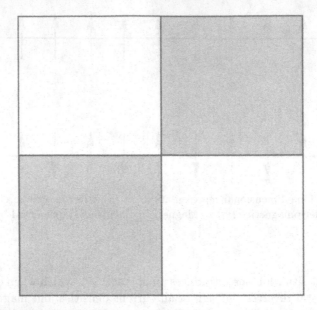

Figure 6.5. This square transforms as $4mm$ if colors are ignored but as the magnetic group $\underline{4mm}$ if colors are included.

Notation and the $\mathcal{M}$ Operation

In describing magnetic groups the international notation (see Table 5.3) is often used. Consider, for example, the group $4mm(C_{4v})$. The eight symmetry operations of this group are $\{E, C_2, 2C_4, 2\sigma_v, 2\sigma_d\}$. Let us examine the operations of this group on a square shaded as shown in Fig. 6.5.

The meaning of magnetic-group notation is that an underlined symbol represents a symmetry operation followed by the $\mathcal{M}$ operation. Thus, each of the fourfold rotation operations as well as one of the two classes of reflections, respectively, are followed by an $\mathcal{M}$ operation. Each such product of operations returns the figure.

Black and White Groups

We wish to obtain the groups which combine the point-group elements and the $\mathcal{M}$ operation. These are called black and white magnetic groups. Let $\mathcal{A} \in G$ be a point symmetry operation and let $\mathcal{M}$ denote the antisymmetry operator and let $L = \mathcal{M}\mathcal{A}$ denote an operator which has the effect of the symmetry operation, $\mathcal{A}$, and the current-reversal operation, $\mathcal{M}$. The element $\mathcal{M}$ is not a member of the black and white group. It follows that both L and $\mathcal{A}$ cannot be members of the black and white group ($\mathcal{M} = L\mathcal{A}^{-1}$ which is not in the magnetic group). To generate the members of the black and white group we partition the group G into the two sets $\{\mathcal{A}_i\}$ and $\{L_k\}$

with the following properties.

$$\mathcal{A}_i \; (i = 1, 2, \ldots, m\} \text{ and } L_k = \mathcal{M}\mathcal{A}_k \; (k = m+1, m+2, \ldots h\} \quad (6.14)$$

where h is the order of the point group G and $S \equiv \{\mathcal{A}_i\}$ is a subgroup of G. Note that $G = \{\mathcal{A}_i\} + \{\mathcal{A}_k\}$ and $G' = \{\mathcal{A}_i\} + \{L_k\}$. The group G is one of the 32 point groups, P_g, discussed previously. Note first that S is an invariant subgroup of G'. This is so because the conjugate of $\mathcal{A}_i$ with any element of G' contains $\mathcal{M}^2 = E$ (conjugate with respect to $\{L_k\}$) or no $\mathcal{M}$ at all (conjugate with respect to S). Furthermore, $\mathcal{A}_k^{-1}\mathcal{A}_i\mathcal{A}_k \in S$, as follows from the rearrangement theorem. Thus S is an invariant subgroup.

We wish to show that G' is a group and that the order of the subgroup S is $h/2$ providing that the order of G' is even. First we establish that

$$\mathcal{A}_k S = \{\mathcal{A}_k\} \quad (6.15a)$$

Assume this is not the case, so that $\mathcal{A}_k\mathcal{A}_i = \mathcal{A}_j$, $(\mathcal{A}_i, \mathcal{A}_j) \in S$ and

$$\mathcal{A}_k\mathcal{A}_i\mathcal{A}_i^{-1} = \mathcal{A}_k = \mathcal{A}_j\mathcal{A}_i^{-1} \in S \quad (6.15b)$$

which is a contradiction. It follows that (6.15a) is valid. In this case,

$$G' = \{\mathcal{A}_i\} + \{L_k\} = S + L_k S = S + \mathcal{M}\mathcal{A}_k S = S + M(G - S) \quad (6.15c)$$

In the last step it is noted that $\mathcal{A}_k S = \{\mathcal{A}_k\} = G - S$. With (6.15a), multiplying the m elements of S by one element of $\{L_k\}$ results in m distinct elements in $\{L_k\}$. With the rearrangement theorem (for h even) we note that multiplying the $(h - m)$ elements of $\{L_k\}$ by an element of $\{L_k\}$ gives $(h - m)$ distinct elements of S. Recapitulating,

$$\mathcal{A}_k S = \{\mathcal{A}_k\} \quad (6.16a)$$

$$\mathcal{A}_k\mathcal{A}_i \in \{L_k\} = M\{\mathcal{A}_k\} \quad (6.16b)$$

$$\mathcal{A}_k\mathcal{A}_{k\prime} \in S \quad (6.16c)$$

It follows that $m = h - m$, or $m = h/2$, so that the invariant subgroup S has half the elements in G'. The relations (6.16) establish closure of the group G'. Furthermore, as each element of S or $\{\mathcal{A}_k\}$ is an element of a point group it has an inverse. It follows that G' is a group and that there are the same number of elements in G' as in G, which establishes the theorem. We may write

$$G' = S \cup \mathcal{M}(G - S) \quad (6.17)$$

This theorem indicates the manner in which a black and white group corresponding to a point group is generated: For any of the 32 point groups G, find the invariant subgroup S. Multiply each element of $G - S$ by $\mathcal{M}$ and join to S to obtain the magnetic group G'. This construction is depicted in Fig. 6.6. Table 6.4 shows 58 black and white groups. The groups C_2, C_3, and T, respectively, do not have subgroups of order $h/2$. The order of the point group is shown in the center column.

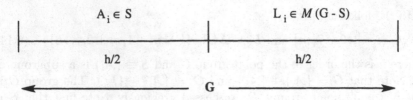

Figure 6.6. Graphical representation of the components of the black and white group $G' = S \cup \mathcal{M}(G - S)$. There are $h/2$ elements in the invariant subgroup, S, and $h/2$ elements in $\mathcal{M}(G - S)$.

Application to C_{2h}

Let us apply this formalism to the $C_{2h} = \{E, C_2, \sigma_h, i\}$ group. The square of each element of this group is E so that there are three invariant subgroups, $(E, C_2), (E, \sigma_h), (E_i)$. In each case the other two elements form the coset which when multiplied by M gives $\{L_k\}$. For example, the magnetic group corresponding to (E, C_2) contains the elements $\{E, C_2, M\sigma_h, Mi\}$.

Table 6.4
Black and White Point Groups*

Ordinary Point Groups				Black and White Point Groups	
	C_1	1	1	None	
	C_i	T	2	$C_i(C_1)$	$\bar{1}$
	C_2	2	2	$C_2(C_1)$	$\underline{2}$
Monoclinic	C_{1h}	m	2	$C_{1h}(C_1)$	$\underline{m}$
	C_{2h}	$2/m$	4	$C_{2h}(C_2)$	$2/\underline{m}$
				$C_{2h}(C_{1h})$	$\underline{2}/m$
				$C_{2h}(C_1)$	$\underline{2}/\underline{m}$
	D_2	222	4	$D_2(C_2)$	$2\underline{22}$
	C_{2v}	$2mm$	4	$C_{2v}(C_2)$	$2\underline{mm}$
				$C_{2v}(C_{1h})$	$\underline{2}mm$
Orthorhombic	D_{2h}	mmm	8	$D_{2h}(D_2)$	$\underline{mmm}$
				$D_{2h}(C_{2v})$	$mm\underline{m}$
				$D_{2h}(C_{2h})$	$\underline{mm}m$
	C_4	4	4	$C_4(C_2)$	$\underline{4}$
	S_4	$\bar{4}$	4	$S_4(C_2)$	$\underline{\bar{4}}$
	D_4	422	8	$D_4(C_4)$	$42\underline{2}$
				$D_4(D_2)$	$\underline{42}$
	C_{4h}	$4/m$	8	$C_{4h}(C_4)$	$4/\underline{m}$
				$C_{4h}(S_4)$	$\underline{4}/\underline{m}$
				$C_{4h}(C_{2h})$	$\underline{4}/m$
Tetragonal	C_{4v}	$4mm$	8	$C_{4v}(C_4)$	$4\underline{mm}$
				$C_{4v}(C_{2v})$	$\underline{4}mm$
	D_{2d}	$\bar{4}2m$	8	$D_{2d}(S_4)$	$\bar{4}2\underline{m}$
				$D_{2d}(D_2)$	$\underline{\bar{4}2}m$
				$D_{2d}(C_{2v})$	$\underline{\bar{4}2m}$
	D_{4h}	$4/mm$	16	$D_{4h}(S_4)$	$4/\underline{mm}$
				$D_{4h}(C_{4v})$	$4/\underline{m}mm$
				$D_{4h}(D_{2h})$	$\underline{4}/mmm$
				$D_{4h}(D_{2d})$	$\underline{4}/\underline{m}mm$
				$D_{4h}(C_{4h})$	$4/m\underline{mm}$

	Ordinary Point Groups			Black and White Point Groups	
Trigonal	C_3	3	3	None	
	D_3	32	6	$D_3(C_3)$	3$\underline{2}$
	C_{3v}	$3m$	6	$C_{3v}(C_3)$	$3\underline{m}$
	S_6	$\bar{3}$	6	$S_6(C_3)$	$\bar{3}$
	D_{3d}	$\bar{3}m$	12	$D_{3d}(S_6)$	$\bar{3}\underline{m}$
				$D_{3d}(C_{3v})$	$\underline{\bar{3}}m$
				$D_{3d}(D_3)$	$\underline{\bar{3}m}$
Hexagonal	C_6	6	6	$C_6(C_3)$	$\underline{6}$
	C_{3h}	$\bar{6}$	6	$C_{3h}(C_3)$	$\underline{\bar{6}}$
	D_{3h}	$\bar{6}m2$	12	$D_{3h}(C_{2h})$	$\bar{6}\underline{m}\underline{2}$
				$D_{3h}(C_{3v})$	$\underline{\bar{6}}m\underline{2}$
				$D_{3h}(C_3)$	$\underline{\bar{6}m2}$
	D_6	622	12	$D_6(C_6)$	$6\underline{2}$
				$D_6(D_3)$	$\underline{6}2$
	C_{6h}	$6/m$	12	$C_{6h}(C_6)$	$6/\underline{m}$
				$C_{6h}(S_6)$	$\underline{6}/\underline{m}$
				$C_{6h}(C_{3h})$	$\underline{6}/m$
	C_{6v}	$6mm$	12	$C_{6v}(C_6)$	$6\underline{mm}$
				$C_{6v}(C_{3v})$	$\underline{6mm}$
	D_{6h}	$6/mmm$	24	$D_{6h}(D_{3h})$	$6/\underline{m}mm$
				$D_{6h}(D_{3d})$	$\underline{6}/\underline{m}mm$
				$D_{6h}(D_6)$	$6/\underline{mm}m$
				$D_{6h}(C_{6v})$	$6/\underline{m}mm$
				$D_{6h}(C_{6h})$	$6/m\underline{mm}$
Cubic	T	23	12	None	
	T_h	$n3$	24	$T_d(T)$	$\underline{m}3$
	T_d	$\bar{4}3m$	24	$T_d(T)$	$\underline{\bar{4}}3m$
	0	432	24	$O(T)$	$\underline{\bar{4}}3$
	0_h	$m3m$	48	$O_h(0)$	$m3\underline{m}$
				$O_h(T_d)$	$m3m$
				$O_h(T_h)$	$m3\underline{m}$

*Black and white magnetic point groups in Schonflies and international notations. An underlined term indicates that it is followed by the time-reversal $\mathcal{M}$ operator. For the Schonflies notation the invariant subgroup is put in parentheses. From: G. Burns, *Introduction to Group Theory with Applications*, Copyright (1977, Academic Press). Reprinted by permission of Academic Press, Inc.)

6.5   Tensors in Group Theory

The GL(n) Group

Tensors were described briefly in Section 6.4. In this section we return to this topic and begin with the concept of the general linear group.

The symbol $\mathbb{R}^n$ denotes an n-dimensional Euclidean space. Let **a** be an $n \times n$ square matrix and let $\mathbf{x}, \mathbf{y}$ be any two vectors in this space and α be a real number. Then **a** is a linear transformation iff

$$\mathbf{a}(\mathbf{x} + \mathbf{y}) = \mathbf{ax} + \mathbf{ay} \tag{6.17b}$$

$$\mathbf{a}(\alpha\mathbf{x}) = \alpha\mathbf{a}(\mathbf{x}) \tag{6.17c}$$

The set of all non-singular linear transformations in $\mathbb{R}^n$ comprise the general linear group of dimension n, $GL(n)$. We recall that a matrix is non-singular providing it has an inverse (Section 3.1). This set of transformations has the following group properties:

(i) If **a** and **b** are non-singular, so is their product, $\mathbf{c} = \mathbf{ab}$.

(ii) Multiplication is associative,

$$\mathbf{a}(\mathbf{bc}) = \mathbf{a}(\mathbf{bc})$$

(iii) The identity **e** and inverse of any element of $GL(n)$ exists,

$$\mathbf{a}^{-1}\mathbf{a} = \mathbf{aa}^{-1} = \mathbf{e}$$

In tensor notation, the transformation $\mathbf{x}' = \mathbf{ax}$ is written (in Einstein convention),

$$x'_j = a_{jk}x_k \tag{6.18a}$$

Tensors of GL(n)

Now consider the n^2 terms $x_i y_j$ formed by taking the products of the components of **x** and **y** in $\mathbb{R}^n$. Applying the transformation (6.18a) to the product forms $x_i y_j$ gives

$$x'_i x'_j = a_{ik}a_{jp}x_k x_p \tag{6.18b}$$

Relabeling $x_i y_j \equiv F_{ij}$ permits the preceding equation to be written

$$F'_{ij} = a_{ik}a_{jp}F_{kp} \tag{6.18c}$$

The form F_{ij} has n^2 components. The values of (i, j) each run from 1 to n. It follows that F_{ij} is a second-rank tensor in n dimensions. Unit basis tensors in this space may be constructed by choosing

$$x_i = \delta_{ik}, \;\; y_p = \delta_{pq} \tag{6.19a}$$

whose products give

$$\Lambda_{ik}^{pq} = \delta_{ip}\delta_{kq} \tag{6.19b}$$

A number of n^2 independent second-rank tensors are obtained by varying p, q over the values, $1, 2, \ldots, n$. The second-rank tensors F_{ij} may be written as linear combinations of the basis tensors, Λ_{ik}^{pq}. It is noted that the tensors F_{ij} are defined with respect to the group $GL(n)$ as they stem from the relation (6.18a).

The third-rank tensor, F_{ijk}, is obtained in similar manner and obeys the transformation rule

$$F'_{ijk} = a_{ir}a_{jp}a_{kq}F_{rpq} \tag{6.20a}$$

Tensors of integer rank, $r > 3$, may be similarly constructed.

Irreducible Tensors

The construction of irreducible tensors with respect to the $GL(n)$ group follows from the construction of product representations of definite symmetry. For example, for tensors of rank 2, one has the symmetric S and antisymmetric A representations

$$F_S = F_{ij} + F_{ji}; \quad F_A = F_{ij} - F_{ji} \tag{6.20b}$$

For third-rank tensors there are three symmetry classes corresponding to the three Young diagrams (Section 4.6):

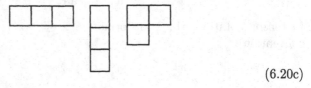

$$(6.20c)$$

relevant to the indices ijk. Symmetrization of tensors of higher rank, with respect to the $GL(n)$ group, may be similarly constructed with corresponding larger sets of Young diagrams which, in accord with rules described in Section 4.6, also correspond to an increase in the number of irreps of the $GL(n)$ group. As the matrix elements a_{ij} are nonrestrictive, reduction of the corresponding tensor space is through the symmetrization process described above. It follows that these symmetrized tensors are irreducible tensors with respect to $GL(n)$. The rth-rank tensors of given symmetry form the basis for an irreducible representation of the $GL(n)$ group.

Dimensionality of Irreps of GL(n)

The dimensionality of an irrep of $GL(n)$ is obtained by constructing the Young tableaux corresponding to the partition of r in n dimensions. For example, the dimensionality of the irrep corresponding to the symmetric tensor of rank r is equal to the number of ways of selecting r different items from the set $2, 3, \ldots, (n+r-1)$ (in row Young diagrams) which is equal to

$$D_S(n,r) = \left(\begin{array}{c} n+r-1 \\ r \end{array} \right) \tag{6.21a}$$

This is the dimensionality of the irrep of $GL(n)$ spanned by the symmetric tensors of rank r. For example, for $n = 2, r = 3$, the integers i, j, k are restricted to $1, 2$ which give the four sets $111, 112, 122, 222$, which agree with (6.21a), $\binom{4}{3} = 4$. For the antisymmetric case (with column Young diagrams) for $r = 3$, in each diagram with three integers from the set of n integers, all three indices must be unequal. It follows that the dimensionality of the irrep of $GL(n)$ spanned by the antisymmetric tensors of rank 3 is $\binom{n}{3}$. Similar reasoning indicates that the dimensionality of the irrep of $GL(n)$, spanned by the antisymmetric tensors of rank $r \leq n$, is

$$D_A(n,r) = \left(\begin{array}{c} n \\ r \end{array} \right) \tag{6.21b}$$

The dimensionality of the irrep of $GL(n)$ spanned by the tensors of mixed symmetry corresponding to the diagram on the right of (6.20c) $(r = 3)$ is given by

$$D_M(n,3) = \frac{n(n^{2-1})}{3} \tag{6.21c}$$

As noted above, in any dimension n, increasing the rank of tensors derived from $GL(n)$ increases the number of related Young diagrams and, correspondingly, the number of irreps of $GL(n)$. It follows that there is a denumerable infinite number of irreps of $GL(n)$.

Regarding antisymmetric combinations of tensors with $r \geq n$, the following property emerges: Consider the case $r = 3, n = 2$ and the antisymmetric form corresponding to F_{ijk} for i, j, k values equal to 1 or 2. Since two indices are equal for this tensor, the antisymmetric combination corresponding to this form must vanish. A similar argument reveals that the antisymmetric combination of tensors of rank $r \geq n$ do not exist.[5] (See, for example, the antisymmetric state for three particles (4.47e). Setting, say, $2 \rightarrow 3$, in this expression, reduces the function to zero.)

---

[5] A similar situation occurs in quantum mechanics. The coupled spin states of three electrons are combinations of the tensor forms $F_{ijk} = \alpha_i \beta_j \gamma_k$ where α_i is, say, the spin state of the 'α' electron and $(i, j, k) = 1, 2$. So F_{ijk} is a tensor with $r = 3$ and $n = 2$. It is known that antisymmetric coupled spin states of three or more electrons do not exist. [See, R.L. Liboff, *Am. J. Physics* **52**, 561 (1984).]

Summary of Topics for Chapter 6

1. Central-field approximation.

2. Atomic notation.

3. Atoms in crystal fields.

4. Reduction of symmetry.

5. Application to a crystal with octahedral symmetry.

6. LS coupling.

7. Splitting of electron levels in crystal fields.

8. Correlation diagrams.

9. Correlation tables.

10. Tensor relations.

11. Electric and magnetic permeabilities.

12. Neumann's principle.

13. Polarizability.

14. Piezoelectric effect.

15. Magnetic classifications.

16. Magnetic groups.

17. Grey, black and white groups.

18. Time-reversal operator.

19. Application to C_{2h}.

20. The general linear group.

21. Tensors in group theory.

Problems

6.1 An atom is placed in a crystal with icosahedral symmetry. Consider an h electron of the atom ($\ell = 5$).
(a) What is the degeneracy of this state?
(b) Into what degeneracies is the ℓ-degeneracy reduced by the crystal field? What are the irreps corresponding to these reduced degeneracies?

6.2 Consider a square lattice in two dimensions that has one of the following two sets of electric dipole moments about each lattice point:

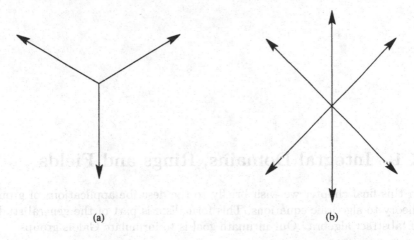

(a)   (b)

Which of these crystals is piezoelectric? Explain your answer.

6.3 In Fig. 6.5: (a) Identify the two classes of reflections. (b) Which class of reflections requires the time-reversal operator for invariance of the colored square?

6.4 Construct the elements of the magnetic groups corresponding to the point group C_{2h}.

6.5 Consider a magnetic crystal with symmetries categorized as 'grey.'
(a) If the order of the uncolored group is h, what is the order of the extended grey group? (b) What effect does application of the time-reversal operator, M, have on the grey group?

7
Elements of Abstract Algebra and the Galois Group

7.1   Integral Domains, Rings and Fields

In this final chapter we wish briefly to the describe applications of group theory to algebraic equations. This formalism is part of the general study of 'abstract algebra.' Our ultimate goal is to formulate Galois groups.

Integral Domains

We start with the definition of an *integral domain* which describes the set of all real and all imaginary numbers. Let D be a set of elements, $a, b, c, \ldots$ for which the sum, $a+b$, and the product, ab, of any two elements, a and b, of D are defined. This domain is an integral domain provided the following nine properties hold:

(a) *Closure.* If $a, b \in D$, then the sum $a+b \in D$ and the product $(ab) \in D$.

(b) *Commutative rules* (addition and multiplication). For all $(a, b) \in D, a + b = b + a, ab = ba$.

(c) *Uniqueness.* If $a = a' \in D$ and $b = b' \in D$, then $a + b = a' + b'$ and $ab = a'b'$.

(d) *Associative rules.* For all $a, b, c \in D, a + (b + c) = (a + b) + c, a(bc) = (ab)c$.

(e) *Distributive rule.* For all $a, b, c \in D, a(b + c) = ab + ac$.

(f) *Existence of zero.* D contains an element 0 such that $a + 0 = a$ for all $a \in D$.

(g) *Existence of unity.* D contains an element $1 \neq 0$ such that $a1 = 1a = a$ for all $a \in D$.

(h) *Existence of additive inverse.* For each $a \in D$, the equation, $a + x = 0$, has a solution, $x \in D$.

(i) *Cancellation rule.* If $c \neq 0$ and $ca = cb$, then $a = b$.

The condition $1 \neq 0$ in (g) precludes the single element 0 from comprising an integral domain.

The integers, rationales, real and complex numbers and modular arithmetics with prime moduli are examples of integral domains. Let $\mathbb{Z}$ denote the set of integers. (zahlen $=$ numbers). The set $\mathbb{Z}(\sqrt{3})$ is the set of all numbers $(a + b\sqrt{3})$ where $a, b \in \mathbb{Z}$. This set is an integral domain (see Problem 7.1). Note that 0 and 1 are identity elements for addition and multiplication, respectively.

Implication of Domain Postulates

Here are five rules derivable from the preceding postulates, relevant to elements of an integral domain.

(1D) If $z \in D$ is such that $a + z = a$ for all $a \in D$, then $z = 0$.

(2D) $a + b = a + c \Longrightarrow b = c$.

(3D) For each $a, b \in D$ there exists one and only one $x \in D$ such that $a + x = b$.

(4D) For all $a, b \in D$, $(-a)(-b) = ab$, where $-a$ is a solution to $a + x = 0$.

(5D) If $ab = 0$, then either $a = 0$ or $b = 0$. (Domain has no divisors of zero.)

If a and b are non-zero elements of D such that $ab = 0$, then a and b are *divisors of zero*. Having defined integral domains, we turn to the closely allied concept of a ring.

Rings

A ring R is a non-empty set that includes the two operations $+$ and $\cdot$ such that:

(i) $a, b \in R \Longrightarrow a + b \in R$.

(ii) $a + b = b + a$ for $a, b \in R$.

(iii) $(a + b) + c = a + (b + c)$ for $a, b, c \in R$.

(iv) An element $0 \in R$ exists such that $a + 0 = a$ for all $a \in R$.

(v) For $a \in R$, there exists $b \in R$ such that $a + b = 0$.

(vi) $a, b \in R \Longrightarrow a \cdot b \in R$.

(vii) $a \cdot (b \cdot c) = (a \cdot b) \cdot c$ for $a, b, c \in R$.

(viii) $a \cdot (b + c) = (a \cdot b) + a \cdot c, \quad (b + c) \cdot a = b \cdot a + c \cdot a \quad a, b, c \in R$.

The condition (ii) indicates that R is an Abelian group under addition. Rings with the condition (vii) are called *associative rings*, which are the rings referred to in this work. None of these eight conditions imply an identity element for multiplication, such that $a \cdot 1 = 1 \cdot a$ for all $a \in R$. When this element is included in the ring, R is called *a ring with identity*. Additionally, these conditions do not imply that $a \cdot b = b \cdot a$. When this commutation property is included, R is called a *commutative ring*.

It is noted that the conditions (i)–(viii) are very similar to those for an integral domain, conditions ($1D$–$5D$). Note that a commutative ring is an integral domain providing it contains an identity element and no divisors of zero. A ring is the basic component whose generalizations give rise to the notions of both an integral domain and a field (defined below).

The sets $\mathbb{Z}$ (integers), $\mathbb{Q}$ (rationales), $\mathbb{R}$ (reals), $\mathbb{A}$ (algebraic numbers) and $\mathbb{C}$ (complex numbers) are all rings. [The group property of integers under addition is described in (1.3).] The set of integers modulo m, $\mathbb{Z}_m$, is a ring as is $R[x]$, the ring of polynomials with coefficients in the commutative ring R. (The concept of modulo is discussed below.) The set of all $n \times n$ matrices comprises a noncommutative ring. Note that for the ring of rationales, for each number $a \neq 0$, there exists a reciprocal or multiplicative inverse $1/a$ such that $(1/a) \cdot a = 1$. However, this is not true for the ring of integers (Problem 1.17).

Fields

Note that the postulates of an integral domain or that of a ring do not include multiplicative inverses for multiplication. This observation gives rise to the concept of a field. A field is an integral domain, D, that contains for each element $a \neq 0$, an identity, e, and an inverse, $a^{-1} \in D$, for which $a^{-1}a = e$. With the closure property of an integral domain, (a), we may conclude that the elements of a field comprise a group under multiplication. The sets $\mathbb{Q}$, $\mathbb{R}$ and $\mathbb{C}$ are fields. A *subfield* of a given field, F, is a subset of F which is itself a field under the operations of addition and multiplication in F. For example, the field of reals is a subfield of the field of complex numbers, $\mathbb{R} \subset \mathbb{C}$. Likewise, $\mathbb{Q} \subset \mathbb{A}$. Every field is an integral domain and every integral domain is a ring (Fig. 7.1).

Isomorphisms and Automorphisms

Two integral domains D and D' are isomorphic iff there exists a one-to-one correspondence between elements $a, b \in D$ and $a', b' \in D'$, such that for

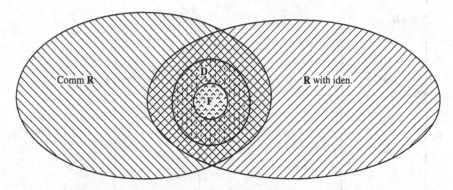

Figure 7.1. Venn diagram illustrating that every field, F, is an integral domain, D, and every integral domain is a ring, R. The left lobe contains commutative rings and the right lobe, rings with identity.

all such elements $(a + b)' = a' + b'$ and $(ab)' = a'b'$. An isomorphism also preserves differences (see Problem 7.2). An isomorphism is written $a \leftrightarrow a'$.

An isomorphism between an integral domain and itself is called an *automorphism*. As an example consider the domain $\mathbb{Z}(\sqrt{3})$. This domain is automorphic to itself under the correspondence $a + b\sqrt{3} \leftrightarrow a - b\sqrt{3}$. (Corresponding properties of point-group mappings were discussed in Section 1.3.)

Homomorphism

For two rings, $a \in R$ and $a' \in R'$, the correspondence $a \to Ha = a'$ is a homomorphism H of R to R' if:

(i) Ha is a uniquely defined element of R' for each element $a \in R$.

(ii) Every element $a' \in R'$ is the image of $a' = Ha$, of at least one $a \in R$.

(iii) For all $a, b \in R$,

$$H(a + b) = Ha + Hb, H(ab) = (Ha)(Hb) \qquad (7.1b)$$

We note that a homomorphism is a correspondence that preserves sums and products. A homomorphism from the ring R to R' includes the additive group of R to that of R' and one may write

$$H0 = 0; \; H(-a) = -Ha, \; H(a - b) = Ha - Hb \qquad (7.1c)$$

where $0'$ is the zero element of the ring R' (the identity element of the additive group of R').

Injective, Surjective and Bijective Maps

Consider the map $f : A \to B$, that maps the set A to the set B. Let $a_1 \in A$ and $a_2 \in B$. Then this map is *injective*, or *one-to-one*, if whenever

A into B

A onto B

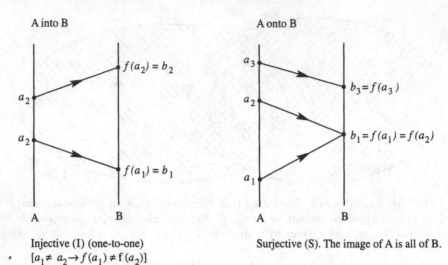

$f(a_2) = b_2$

a_2

a_2

$f(a_1) = b_1$

A   B

a_3

$b_3 = f(a_3)$

a_2

$b_1 = f(a_1) = f(a_2)$

a_1

A   B

Injective (I) (one-to-one)

$[a_1 \neq a_2 \rightarrow f(a_1) \neq f(a_2)]$

Surjective (S). The image of A is all of B.

A onto B

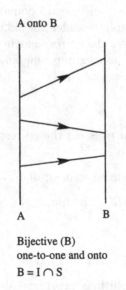

A   B

Bijective (B)

one-to-one and onto

$B = I \cap S$

Figure 7.2. Onto and into maps. Arrows indicate the direction of mappings.

$a_1 \neq a_2$, then $f(a_1) \neq f(a_2)$. The map f is *surjective*, or *onto*, if for each $b \in B$ there exists some $a \in A$ such that $f(a) = b$. That is, the image of A is all of B. The mapping f is *bijective* if it is both one-to-one and onto. Equivalently, one may say, $A \rightarrow B$ is *one-to-one* if each $b \in B$ has at most one $a \in A$. The mapping $A \rightarrow B$ is *onto* if each $b \in B$ has at least one $a \in A$ that maps onto b (Fig. 7.2). Note that a bijective map is the intersection of an injective (I) and a surjective (S) map. Here are some examples: The mapping $f(z) = \text{Re}(z)$, where $z \in \mathbb{Z}$ maps $\mathbb{C}$ onto $\mathbb{R}$, the set of real numbers, and is surjective. The linear transformation $f(x) = ax + b$, where a and b are real numbers and $x \in \mathbb{R}$ maps $\mathbb{R}$ onto $\mathbb{R}$, is one-to-one and onto, and thus is bijective.

Ideal of a Ring

The notion of cosets and invariant subgroups were described in Section 5.1. Similar concepts exist for rings. An *ideal* C in a ring R is a subgroup of the additive group of R. Each element $a \in R$ belongs to a coset $a' = a + C$, consisting of the sums, $a + c$, for $c \in C$. The subgroup C is an invariant subgroup of the additive group R. An ideal C in a ring R has the two defining properties:

$$\text{(i)} \quad (c_1, c_2) \in C \Rightarrow (c_1 - c_2) \in C \qquad (7.1d)$$

$$\text{(ii)} \quad c \in C \text{ and } a \in R \Rightarrow ac \text{ and } ca \in C \qquad (7.1e)$$

One then has the

Theorem. In any homomorphism H of a ring, the set of elements mapped onto $0'$ is an ideal in R.

Proof: Let $c \in R$ for which $Hc = 0'$. Then for $a \in R$, $H(ac) = (Ha)(Hc) = (Ha)0' = 0$. Likewise $H(ca) = 0'$, which establishes (ii). The values $Hc_1 - Hc_2 = 0'$, with (7.1c), give

$$H(c_1 - c_2) = Hc_1 - Hc_2 = 0' - 0' = 0'$$

which establishes property (i). The ideal C mapped into $0'$ of R' in the kernel of the homomorphism H (see Section 1.3). Since C is an additive subgroup of R, the quotient group, R/C, exists. This quotient group is the set of all cosets, $a + C$, as a runs over R. Let us show that the only ideals a field F has are (0) and the field F itself. Suppose that $C \neq (0)$ is an ideal of F. Let $a \neq 0 \in C$. As C is an ideal of F, $1 = a^{-1}a \in C$. Since $1 \in C, r1 = r \in C$, for every $r \in F$ [by (7.1e)] so that $C = F$. It follows that F has only the trivial ideals (0) and F itself.

7.2 Numbers

Integers, rationales and irrationals comprise the set of real numbers. Integers $(0, \pm 1, \pm 2, \ldots)$ may be further divided into positive and negative integers, primes, and non-primes, and even and odd numbers. If a and b are real numbers, then $a + ib$ is a complex number, where $i = \sqrt{-1}$.

Another set of numbers are the *algebraic numbers*. An algebraic number, x, is a complex number which satisfies a polynomial equation, with rational coefficients not all zero, of the form

$$a_0 + a_1 x + a_2 x^2 + \cdots + a_n x^n = 0 \qquad (7.2)$$

A theorem due to Cantor[1] states that the set of algebraic numbers is countable. It follows that the set of real numbers is larger than the set of algebraic numbers. The set of algebraic numbers forms the field $\mathbb{A}$.

Numbers which do not satisfy a polynomial equation with rational coefficients are called *transcendental* numbers. Examples include π and $e = 2.71828\ldots$. On the other hand, numbers such as $5^{3/2}, \sqrt{-3}$, etc., are algebraic.

Congruences and modulo

The phrase

$$a \equiv b \pmod{m} \tag{7.3a}$$

means that

$$a = b + k \cdot m \tag{7.3b}$$

where k is an integer and $m \neq 0$. That is, the difference $a - b$ is divisible by m. The relation (7.3a) is described as 'a is congruent to b, modulo m.'[2] For example, the angle subtended on a circle, θ, is measured with respect to 2π, so that $\theta = \theta_0 \pmod{2\pi}$ means that θ and θ_0 are separated by an integral multiple of 2π. Here are some other simple examples of this concept,

$$\begin{array}{ll} 20 \equiv 10 \pmod{5}, & 20 = 10 + 2 \cdot 5 \\ -11 \equiv 5 \pmod{8}, & -11 = 5 - 2 \cdot 8 \\ 18 \equiv 0 \pmod{9}, & 18 = 0 + 2 \cdot 9 \end{array}$$

With respect to integers (a, b), one notes that $a \equiv b \pmod{m}$ iff $m|(a - b)$. The symbol $a|b$ means that a divides b.

Fermat's little theorem states that if a is an integer and p is a prime and p does not divide a, then

$$a^p \equiv a \pmod{p} \tag{7.3c}$$

Here is a proof: For fixed prime p, let $P(n)$ be the proposition that $n^p \equiv n \pmod{p}$. If this proposition can be proved for all integer n, then Fermat's theorem is established. We demonstrate its validity by mathematical induction. This proposition is evidently valid for $n = 0$ and $n = 1$. In the binomial expansion of $(n + 1)^p$, every coefficient, $p!/k!(p - k)!$, except the first and last, are divisible by p. We recall that

$$(n + 1)^p = n^p + \cdots + \binom{p}{k} n^{p-k} + \cdots + 1 \tag{7.3d}$$

---

[1] Georg Cantor (1845–1919).
[2] The notion of congruences is due to Carl F. Gauss (1777–1855).

where k is an integer, $0 \leq k \leq p$, and $\binom{p}{k}$ is integer. The preceding may be rewritten

$$(n+1)^p = n^p + 1 \ldots \tag{7.3e}$$

$$(n+1)^p \equiv (n^p + 1)(\operatorname{mod} p) \tag{7.3f}$$

The later equality stems from the fact that the terms on the right of (7.3d) minus $(n^p + 1)$ are all integer multiples of p. With $P(n)$, the preceding may be written

$$(n+1)^p \equiv (n+1) \ (\operatorname{mod} p) \tag{7.3g}$$

It follows that $P(n) \Rightarrow P(n+1)$. QED.

A theorem of Wilson[3] states that

$$(p-1)! \equiv -1 \ (\operatorname{mod} p), \tag{7.3h}$$

so that $p \mid [(p-1)! + 1]$ where p is prime.

Elementary Congruences

Here are four elementary congruences. With (7.3a,b) we write

$$a \equiv b \ (\operatorname{mod} m) \Rightarrow b \equiv a \ (\operatorname{mod} m) \tag{7.4a}$$

With the first relation we write $b = a - km$. Our next congruence relation states that if

$$a \equiv b \ (\operatorname{mod} m), b \equiv c \ (\operatorname{mod} m), \tag{7.4b}$$

then

$$a \equiv c \ (\operatorname{mod} m) \tag{7.4c}$$

The first relation in (7.4b) implies that $a - b = mk$ and the second that $b - c = mh$, so that $a - c = m(k+h)$ which is the statement (7.4c).

Our second congruence indicates that if a and b are any two integers, then

$$a \equiv b \ (\operatorname{mod} 1) \tag{7.4d}$$

as $a - b = k \cdot 1 = k =$ an integer. Next we note that

$$a \equiv b \ (\operatorname{mod} m) \Rightarrow ak \equiv bk \ (\operatorname{mod} m), a^k \equiv b^k \ (\operatorname{mod} m) \tag{7.4e}$$

for all integer k. With the latter equality and (7.5d) we find $(a-b) \mid (a^k - b^k)$ [note that $a = b$ is a root of both sides of this relation] and $m \mid (a-b)$. It follows that $m \mid (a^k - b^k)$.

Lastly, we note that (see Problem 7.13)

$$a \equiv b \ (\operatorname{mod} m) \Rightarrow (a+c) \equiv (b+c) \ (\operatorname{mod} m) \tag{7.4f}$$

---

[3]Sir John Wilson (1734–1793).

Algebra of Congruences

Addition

Consider the congruences,

$$a = b \ (\text{mod} \, m), \ c = d \ (\text{mod} \, m) \tag{7.5a}$$

These imply that

$$a + c = (b + d) \ (\text{mod} \, m) \tag{7.5b}$$

Subtraction

From the preceding, it follows that

$$a - c \equiv (b - d) \ (\text{mod} \, m) \tag{7.5c}$$

Multiplication

Multiplying the components (7.5a) gives, with the preceding definitions,

$$ac \equiv bd + m(kd + bh + mkh)$$

so that

$$ac \equiv bd \ (\text{mod} \, m); \ \ a^2 \equiv b^2 \ (\text{mod} \, m); \ \ a^k \equiv b^k \ (\text{mod} \, m) \tag{7.5d}$$

Cancellation

We wish to discover the condition under which

$$ab \equiv 0 \ (\text{mod} \, n) \Rightarrow a \equiv 0 \ (\text{mod} \, n) \text{ or } b \equiv 0 \ (\text{mod} \, n) \tag{7.5e}$$

These relations are equivalent to the properties $n|ab \Rightarrow n|a$ or $n|b$, which is valid provided n is prime. Let $\mathbb{Z}_n$ denote the set of integers $(\text{mod} \, n)$. One then has the

Theorem. The cancellation law for multiplication holds for the set $\mathbb{Z}_n$ iff n is prime.

Residue Classes

For fixed (a, m) the arithmetic progression $x = a + mk$ is called the *residue class* $(\text{mod} \, m)$ denoted by $\bar{a}$. At a given value m, the set of integers $\mathbb{Z}$ is partitioned into m residue classes, $\bar{0}, \bar{1}, \ldots, \overline{m-1}$. For example, at $m = 3$,

$$\bar{0} = \{0 + 3k\}, \ \bar{1} = \{1 + 3k\}, \ \bar{2} = \{2 + 3k\}$$

In the set of integers corresponding to r $(\text{mod} \, n)$, each integer belongs to one and only one residue class. Two integers belong to the same residue class iff they are congruent.

Field Property of Residue Classes

With (7.5d) we see that $a \pmod{m}$ satisfies closure with respect to multiplication. The identity element in this case is

$$E \equiv 1 \pmod{m} \tag{7.6a}$$

The elements of the *residue class* $\pmod{p}$, where p is prime, have inverses and form a field. For example, consider the residue class $\pmod{5}$: $\bar{0}, \bar{1}, \bar{2}, \bar{3}, \bar{4}$. Omitting $\bar{0}$, these elements comprise a field, with the following group table, with $E = \bar{1}$:

	$\bar{1}$	$\bar{2}$	$\bar{3}$	$\bar{4}$
$\bar{1}$	$\bar{1}$	$\bar{2}$	$\bar{3}$	$\bar{4}$
$\bar{2}$	$\bar{2}$	$\bar{4}$	$\bar{1}$	$\bar{3}$
$\bar{3}$	$\bar{3}$	$\bar{1}$	$\bar{4}$	$\bar{2}$
$\bar{4}$	$\bar{4}$	$\bar{3}$	$\bar{2}$	$\bar{1}$

$$\tag{7.6b}$$

For example, consider the product

$$\bar{2} \cdot \bar{3} = (2 + 5k) \cdot (3 + 5k) = 6 + 5k' = 1 + 5k'' = \bar{1}$$

so that, for example, $\bar{2}^{\,-1} = \bar{3}$. In this manner one finds that the residue class $\pmod{p}$, omitting $\bar{0}$, is a field.

If the sum and product of residue classes are defined by the equations

$$\bar{a} + \bar{c} = \overline{a + c}, \quad \bar{a} \cdot \bar{c} = \overline{a \cdot c} \tag{7.6c}$$

then the set of *residue classes* $\pmod{m}$, $\mathbb{Z}_m$, is a ring. The mapping $\phi a = \bar{a}$ is a homomorphism of $\mathbb{Z}$ onto $\mathbb{Z}_m$.

Greatest Common Divisor

The greatest common divisor (gcd) of two numbers a and b is written (a, b). For example, $(4, 48) = 4$. If a and b have only the divisor 1, then a and b are said to be *relatively prime*. For example, $(3, 10) = 1$. We note that the gcd of two integers, a, b may be written as a linear combination of a and b, namely,

$$(a, b) = sa + tb \tag{7.7a}$$

where s and t are integers. For example,

$$(10, 35) = 5 = s \cdot 10 + t \cdot 35$$
$$= -3 \cdot 10 + 1 \cdot 35$$

Here are four theorems regarding gcd:

I. If $a|b$, then

$$b \equiv 0 \pmod{a} \tag{7.7b}$$

We note that $0 \pmod{a} = ka = b$.

II. If

$$(a, c) = 1 \text{ and } c|ab, \quad \text{then } c|b \qquad (7.7c)$$

To prove this theorem we note that since $(a, c) = 1$, with (7.7c),

$$1 = sa + tc$$

Multiplying both sides by b gives

$$b = sab + tcb$$

As c divides both terms on the right we conclude that $c|b$.

III. If

$$ac \pmod{m} \equiv bc \pmod{m} \text{ and } (c, m) = 1 \qquad (7.7d)$$

then c may be canceled from both sides of the first equation. To prove this theorem, we note that the left equality in (7.7d) implies

$$c(b - a) = km$$

so that $m|c(b-a)$. But $(m, c) = 1$ and, by the preceding theorem, $m|(b-a)$ so we may write

$$a \equiv b \pmod{m} \qquad (7.7e)$$

It follows that for congruences such as (7.7d) one may include division in the preceding list of algebraic relations.

IV. If $a|c$ and $b|c$ and $(a, b) = 1$, then $ab|c$. To obtain this result, with the given conditions, we write $c = ar = bs$ (all integers). With (7.7b), we write $1 = ax + by$, where x, y are also integers. Multiplying by c gives $c = cax + cby = (bs)ax + (ar)by$, which proves the theorem.

The Chinese Remainder Theorem

This theorem states the following. Consider the sequence $(n_1, n_2, \ldots, n_r)$ of relatively prime integers and let $\{a_i\}$ be arbitrary constants. Then the system of congruences

$$x \equiv a_1 \pmod{n_1}, x \equiv a_2 \pmod{n_2}, \ldots, x \equiv a_r \pmod{n_r} \qquad (7.8a)$$

has a simultaneous solution which is unique modulo the integer,

$$n = n_1 n_2 \ldots n_r.$$

To prove this theorem, we set

$$N_k = n_1 \ldots n_{k-1} n_{k+1} \ldots n_r \qquad (7.8b)$$

which is the product of n_i factors with n_k missing. Let x_k be the unique solution to the congruence $N_k x \equiv 1 \pmod{n_k}$. We wish them to show that the integer

$$\tilde{x} = a_1 N_1 x_1 + a_2 N_2 x_2 + \ldots + a_r N_r x_r \tag{7.8c}$$

is a simultaneous solution of (7.8a). Note that $N_i \equiv 0 \pmod{n_k}$ for $i \neq k$, since $n_k | N_i$. The preceding equation reduces to

$$\tilde{x} = a_1 N_1 x_1 + a_2 N_2 x_2 + \ldots + a_r N_r x_r \equiv a_k N_k x_k \pmod{n_k} \tag{7.8d}$$

and we may write

$$\tilde{x} \equiv a_k \cdot 1 \equiv a_k \pmod{n_k}, \ k = 1, 2, \ldots, r \tag{7.8e}$$

which are the relations (7.8a). To show the uniqueness of (7.8a), we assume that x' is another integer which satisfies these congruences, in which case

$$\tilde{x} \equiv a_k \equiv x' \pmod{n_k}, k = 1, 2, \ldots, r \tag{7.8f}$$

so that $n_k | (\tilde{x} - x')$ for each value of k. Combining this relation with the fact that $(n_i, n_j) = 1$ and theorem **IV** above, indicates that $n | (\tilde{x} - x')$, and we may write (with **I**), $\tilde{x} \equiv x' \pmod{n}$, which establishes the theorem.

7.3 Irreducible Polynomials

Consider the polynomial

$$f(x) = a_0 + a_1 x + a_2 x^2 + \ldots + a_n x^n \tag{7.8g}$$

The *degree* of this polynomial is the integer n (provided that $a_n \neq 0$).

A polynomial is *reducible* over a field F if it can be factored into polynomials of lower degree with coefficients in F; otherwise it is called *irreducible*. The polynomial $x^2 + n^2$, where n is an integer, is irreducible over the field of reals. It may be factored as $(x + in)(x - in)$, where $i = \sqrt{-1}$, which is not in the field of reals. The polynomial $x^2 - p$, where p is a prime number, is irreducible over the field of rationales, as $\sqrt{p}$ is an irrational number.

Eisenstein's Irreducibility Criterion[4]

This criterion is expressed in the following theorem. Let p be prime and let

$$a(x) = a_0 + a_1 x + a_2 x^2 + \ldots + a_n x^n \tag{7.9a}$$

be a polynomial with integral coefficients such that $a_n \neq 0 \pmod{p}$, $a_0 \neq 0 \pmod{p^2}$ and

$$a_0 = a_1 = \ldots = a_{n-1} = 0 \pmod{p} \tag{7.9b}$$

---

[4]Named for F.G. Eisenstein (1823–1852).

(i.e., each coefficient $a_i, i \neq n$, is a multiple of p). Then $a(x)$ is irreducible over the field of rationales.

An example of this theorem is given by the 'cyclotomic polynomial' (whose roots are the p roots of unity)

$$\phi(x) = (x^p - 1)/(x - 1) = x^{p-1} + x^{p-2} + \ldots + x + 1 \qquad (7.9c)$$

This polynomial does not satisfy Eisenstein's criterion. However, the change in variables $y = x - 1$, together with the binomial expansion, gives

$$(x^p - 1)/(x - 1) = [(y + 1)^p - 1]/y$$

$$= y^{p-1} + py^{p-2} + \frac{p(p - 1)}{1 \cdot 2} y^{p-3} + \ldots + p \qquad (7.9d)$$

As all coefficients in this polynomial are divisible by p (i.e., all are congruent to $0 \bmod p$), (7.9d) satisfies Eisenstein's criterion and is therefore irreducible. This establishes the irreducibility of the cyclotomic polynomial (7.9c).

GCD of Two Polynomials. Relatively Prime Functions

Let $a(x) \neq 0$ and $b(x)$ be any two polynomials over the field F. Then one may write

$$b(x) = q(x)a(x) + r(x) \qquad (7.10)$$

where $q(x)$ and $r(x)$ are polynomials over F and $r(x)$ is of degree less than $a(x)$. For example, the polynomials

$$a(x) = (x - 2)^2, \quad b(x) = (x - 2)(x^2 - 6x + 9)$$

give

$$b(x) = (x - 4)a(x) + (x - 2)$$

for which $r(x)$ is of degree less than $a(x)$.

When $r(x) = 0$ in (7.10) one says that $b(x)$ is divisible by $a(x)$, i.e., $a(x)|b(x)$.

Let $F[x]$ denote the domain of polynomial forms in one indeterminate x over the field F. (Equivalently, $F[x]$ denotes a polynomial ring over the field F.) Any two polynomials, $a(x)$ and $b(x)$, in $F[x]$, will have a *gcd* $d(x)$, with the following properties:

(i) $d(x)|a(x)$,

(ii) $d(x)|b(x)$,

(iii) $c(x)|a(x)$ and $c(x)|d(x)$ imply $c(x)|d(x)$,

(iv) $d(x)$ is the linear combination

$$d(x) = s(x)a(x) + t(x)b(x), \qquad (7.11)$$

where s and t are polynomial coefficients [compare (7.11) to (7.7a)].

Two polynomials $a(x)$ and $b(x)$ in $F[x]$ are *relatively prime* iff their only common factors are the constants of $F[x]$. In general, if R is a ring, then two elements of R are relatively prime if their *gcd* is a *unit element* (Problem 7.26). Constants in the ring $F[x]$ are unit elements. Such is the case in (7.10) if q is a number and $r = 0$. In this event, $d = 1$.

An example of (7.11) is given by $a(x) = (x-1)^3$ and $b(x) = (x-1)^2(x-2)$ for which $d(x) = (x-1)^2$ is the linear combination

$$d(x) = [a(x) - b(x)]$$

Extension of a Field

The *extension* K of a field F is any field which contains F as a subfield ($F \subset K$). An extension of F may be generated by elements of F and an external parameter. For example, the field $\mathbb{Q}[x]$ of all rational polynomial forms [left side of (7.2)] is the field $\mathbb{Q}$ adjoined with the element x. A single field may be generated in a number of ways. The field $\mathbb{Q}(\sqrt{2})$ is generated by the $\sqrt{2}$ root of the equation $x^2 - 2 = 0$ and consists of all real numbers $a + b^2\sqrt{2}$ with rational coefficients a, b. The root, $-2 + \sqrt{2}$ (the *generator* of the field), of the equation $x^2 + 4x + 2 = 0$ likewise generates the field, $\mathbb{Q}(\sqrt{2})$. Any number in the field may be expressed in terms of the new generator as

$$a + b\sqrt{2} = (a + 2b) + b(-2 + \sqrt{2})$$

The field $\mathbb{Q}$ is extended by adjoining the roots of the polynomial $x^2 - 2$. We note $\mathbb{R} \subset \mathbb{C}$ and $\mathbb{Q} \subset \mathbb{R}$. That is, the field of complex numbers is an extension of the field of reals and the field of reals is an extension of the field of rationales.

Root Fields

The set composed of conjugate roots of an irreducible polynomial gives rise to the notion of the *root field* defined as follows. An extension N of the field F is a root field of a polynomial $f(x)$ of degree n with coefficients in F if:

(i) $f(x)$ can be factored into a linear factor in N : $f(x) = c(x-u_1) \ldots (x-u_n)$;

(ii) N is generated over the F by the roots of $f(x)$ as $N = F(u_1, \ldots, u_n)$; where c is a unit element. We note the

Theorem. A polynomial over any field has a root field.

A field K is a *simple* extension of the field F iff K is generated by a single element x, so that $K = F(x)$. If x is algebraic, then K is a *simple algebraic* extension of F. The field $\mathbb{Q}(\sqrt{2})$ is a simple algebraic extension of the field of rationales, $\mathbb{Q}$, generated by the element $\sqrt{2}$ which is of degree 2 over the rationales.

Conjugate over a Field

Two elements u and v of a field K are conjugate over a subfield F of K, iff u and v are both roots of the same polynomial, irreducible over F. Consider, for example, the polynomial $x^3 - 5$ over the field $\mathbb{Q}$, which has the root field $\mathbb{Q}(5^{1/3}, \omega 5^{1/3}, \omega^2 5^{1/3})$, where $\omega = \exp(i2\pi/3)$. This field may be generated by the two algebraic numbers ω and $5^{1/3}$, namely, $\mathbb{Q}(5^{1/3}, \omega)$. The number $5^{1/3}$ satisfies a third-order polynomial and has degree 3 over the field $\mathbb{Q}(5^{1/3})$. The factor ω satisfies the cyclotomic equation $x^2 + x + 1 = 0$, which is irreducible over the field of reals. Its roots are both complex and ω has degree 2 over $\mathbb{Q}$ $(5^{1/3})$. The field $\mathbb{Q}$ $(5^{1/3}, \omega)$ is sixth order where a and b are rational numbers. These numbers, which are elements of the field of complex numbers $\mathbb{C}$, are conjugate over $\mathbb{Q}$.

A second example of the extension of a field by roots of a polynomial is as follows. Specifically, consider the extension of the finite field, $\mathbb{Z}_3$, composed of three residue classes of integers 0, 1, 2 (mod 3), and extend it to include the element u which is a root of the polynomial

$$f(u) = u^2 - u - 1.$$

The extension, K, consists of nine elements of the form $a + bu$, where a and b are chosen from the residue classes $0, 1, 2$. Let us verify these nine elements, $a + bu$, for the field properties of addition and multiplication.

Addition

$$(a + bu) + (c + cu) = (a + c) + (b + d)u \qquad (7.12a)$$

Multiplication

$$(a + bu)(c + bd) = (ac + bd) + (ad + bc + db)u \qquad (7.12b)$$

The right sides of the latter two equations are elements of the extension $\mathbb{Z}_3(u)$ of $\mathbb{Z}_3$.

We state the following theorem important to these discussions: For every polynomial $p(x)$, irreducible over a field F, there exists a field which is a simple algebraic extension of F generated by a root u of $p(x)$. For example, $\sqrt{2}$ is a root of the polynomial $x^2 - 1$. The field $\mathbb{Q}(\sqrt{2})$, composed of elements of the form $a + b\sqrt{2}$, where a and b are rationales, is an extension of the field $\mathbb{Q}$. Therefore, $\mathbb{Q}$ is a subfield of $\mathbb{Q}(\sqrt{2})$ and we write $\mathbb{Q} \subset \mathbb{Q}(\sqrt{2})$.

Monic Polynomials

With reference to (7.8), if the coefficient of the term of highest power is unity, the polynomial is called *monic*. A polynomial which is not monic may be made so by multiplying through by the inverse of the constant that multiplies the highest power. The product of two monic polynomials is monic.

Let us prove the following theorem: Any rational root of a monic polynomial with integral coefficients is an integer, r, which has the property, $r|a_n$, where a_n is the constant coefficient in the monic polynomial

$$p(x) = x^n + a_1 x^{n-1} + \ldots + a_n \qquad (7.13a)$$

The coefficients a_i are integers. Consider the rational root r/s where r and s are relatively prime integers, $(r, s) = 1$. Substituting this value into the equation $p(x) = 0$ gives

$$0 = s^n p(r/s) = r^n + a_1 r^{n-1} s + \ldots + a_n s^n \qquad (7.13b)$$

so that

$$-r^n = s[a_1 r^{n-1} + \ldots + a_n s^{n-1}] \qquad (7.13c)$$

and $s|r^n$. But $(r, s) = 1$, and with successive application of Rule 1 (7.7b), we conclude that $s|r^{n-1}$, ..., $s|1$. It follows that $s = 1$ and the rational root, r/s, reduces to r, an integer. Substituting this value in (7.13c) gives

$$-a_n = r[r^{n-1} + \ldots + a_{n-1}] \qquad (7.13d)$$

which indicates that $r|a_n$, establishing the proof.

Elements Algebraic over a Field

Definition. Let K be any field and F any subfield of K. An element y of K is called 'algebraic over F' if y satisfies a polynomial equation with coefficients not all zero in F,

$$a_0 + a_1 y + a_2 y^2 + \ldots + a_n y^n = 0 \qquad (7.14)$$

Theorem. An element u that is algebraic over a field F is the root of exactly one monic polynomial $p(x)$ irreducible in the domain of all polynomials over $F[x]$. The element u is a root of another polynomial $g(x)$ with coefficients in F iff $g(x)$ is a multiple of $p(x)$ in the domain $F[x]$. For example, the polynomial, $x^2 - 3$, is monic and is an element of $\mathbb{Q}[x]$. It is irreducible over the rational field and, apart from a numerical factor, is unique. Up to multiplication by a unit element, over the field $\mathbb{C}$ of complex numbers, all irreducible polynomials are linear. Any such linear component is likewise unique. We recall that if a field, F, is algebraically complete, then every polynomial in $F[x]$ has a root in F. It follows that the irreducible polynomials of an algebraically complete field are linear. In this context, we note the

Fundamental Theorem of Algebra. Any field has an algebraically complete extension.

Employing the preceding theorem, we wish to show that the field generated by F and the algebraic element $u \in K$ is a subfield of K. We recall that each element of a field has an inverse. Let us find the inverse of a non-zero element $f(u)$ in $F[u]$. The statement that $f(u) \neq 0$ means that u is not a root of $f(x)$ so that, by the above theorem, $f(x)$ is not a multiple of the irreducible polynomial $p(x)$. Hence these two functions are relatively prime and with (7.11) one may write

$$1 = t(x)f(x) + s(x)p(x) \tag{7.15}$$

where $t(x)$ and $s(x)$ are elements of $F[x]$. The preceding equation in the field $F[u]$ is $1 = t(u)f(u)$. This relation reveals that the non-zero element $f(u)$ of $F[u]$ has a reciprocal $t(u)$ that is also an element $F[u]$. We conclude that $F[u]$ is a subfield of K.

Definition. The degree, n, of an element u algebraic over a field F is the degree n of the monic irreducible polynomial with coefficients in F and root u and is written $n = [u : F]$. For example, the pth root of unity, which satisfies the cyclotomic equation (7.9c), is algebraic over the field of reals and is of degree $p - 1$.

Theorem. Consider an element u that is algebraic over a field F and is of order n. Then each element of the subfield $F(u)$ generated by F and u can be uniquely represented as a polynomial

$$a_0 + a_1 u + a_2 u^{(2)} + \ldots + a_{n-1} u^{(n-1)} \tag{7.16}$$

with coefficients $\{a_i\}$ in F and of degree at most $n - 1$. To add or subtract two such polynomials, one adds or subtracts the corresponding coefficients. To multiply them, one forms the polynomial product and then computes the remainder of the product after dividing by $p(x)$. Note that (7.16) has the form of an inner product of the vectors $(a_0, a_1, \ldots .)$ and the basis $(1, u^{(1)}, u^{(2)} \ldots)$. In this context we note the

Theorem. The degree of an algebraic element u over a field F is equal to the dimension of the extension $F(u)$ regarded as a vector space over F with basis $(1, u^{(1)}, u^{(2)} \ldots)$.

Subfields

We return to the notion of the generation of a subfield. Specifically let $F \subset K, y \in K$ and coefficients $a_i \in F$. Then the polynomials

$$F(y) = a_0 + a_1 y + \ldots + a_n y^n \tag{7.17}$$

comprise a subdomain of K containing F and y. The set of all such polynomials, $F[y]$, is closed under addition, subtraction and multiplication. A field may be formed from these elements in the following manner. If $f(y) \neq 0$ and $g(y) \neq 0$ are elements of K, then their quotient $f(y)/g(y)$ is the field generated by F and y, and is labeled $F(y)$. Note that in this case, inverses of elements $F(y)$ exist by construction and with preceding properties, $F(y)$ is a subfield of K.

As noted previously, if y does not satisfy the polynomial equation $f(y) = 0$, then y is transcendental over the field F. In this case one has the following

Theorem. If y is transcendental over F, the subfield $F(y)$ generated by F and y is isomorphic to the field $F[x]$ of all rational polynomial forms, in an indeterminate x, with coefficients in F. The isomorphism may be constructed so that $c \leftrightarrow x$ and $z \leftrightarrow a$, for each $a \in F$.

Example

We wish to offer two independent proofs of the

Theorem. The form $p^{1/n}$ is an irrational number, where p is prime and $n \neq 0, 1$ is an integer. Here are three fundamental lemmas for our first proof:

Lemmas

 I. (*Fundamental Theorem of Arithmetic.*) Every integer $q > 1$ can be expressed, apart from order of the elements, as a unique product of primes (see Problem 7.27).

 II. If p is prime and $p|ab$, then $p|a$ or $p|b$.

 III. If p is prime and $p|a_1a_2\ldots a_n$, then $p|a_k$ for some k where $1 \leq k \leq n$.

To establish the desired result, we assume the contrary, namely, that

$$p^{1/n} = a/b \qquad (7.18a)$$

where a, b, n are positive integers and $(a, b) = 1$. Relation (7.18a) gives

$$b^n p = a^n \qquad (7.18b)$$

which implies that $b|a^n$. Lemma I guarantees the existence of a prime p' such that $p'|b$. With Lemma II we find that $p'|a^n$ which, with Lemma III, implies that $p'|a$. It follows that $(a, b) \geq p'$, which is a contradiction to our hypothesis unless $b = 1$. In this case, (7.18b) gives $p = a^n$, which implies that $a|p$ which is a contradiction, so assumption (7.18a) is false. We conclude that $p^{1/n}$ is an irrational number.

The second proof of this theorem stems from the theorem established above: Any rational root of a monic polynomial with integral coefficients is an integer. It follows that the non-integer roots of such a polynomial equation must be irrational. The polynomial equation

$$x^n - p = 0 \qquad (7.18c)$$

satisfies the theorem's criterion. An integral solution, a, to this equation gives $p = a^n$, which is a contradiction. So if the root is not rational, it must be irrational and we recapture the preceding theorem.

7.4   The Galois Group

Automorphism

The concept of the automorphism of a field was described in Section 7.1. We repeat: An automorphism T of a field K is a one-to-one correspondence $a \leftrightarrow Ta$ of the set K with itself such that sums and products are preserved. Thus for all a and b in K,

$$T(a+b) = Ta + Tb, \quad T(ab) = (Ta)(Tb) \tag{7.19}$$

The product ST of two automorphisms is also an automorphism, and the inverse of an automorphism is likewise an automorphism. It follows that: The set of automorphisms of a field K forms a group.

As an example, consider the field $K = \mathbb{Q}(\sqrt{2}, i)$ of degree 4 over the field of rationals generated by $\sqrt{2}$ and i. With respect to the intermediate field $F = \mathbb{Q}(i)$, the full field K is an extension of degree 2, generated by either of the conjugate roots, $\pm\sqrt{2}$, roots of $x^2 - 2$. These roots are algebraically indistinguishable and there is an automorphism S of K which takes $\sqrt{2}$ into $-\sqrt{2}$, with the elements of $\mathbb{Q}(i)$ remaining fixed. The effect of S on an element of K is given by

$$S(a + b\sqrt{2} + ci + d\sqrt{2}i) = a - b\sqrt{2} + ci - d\sqrt{2}i \tag{7.20a}$$

corresponding to the basis $\{1, \sqrt{2}, i, \sqrt{2}i\}$ (Problem 7.18). Similarly, there is an automorphism T leaving the members of $\mathbb{Q}(\sqrt{2})$ invariant and carrying i into $-i$. For this mapping we write

$$T(a + b\sqrt{2} + ci + d\sqrt{2}i) = a + b\sqrt{2} - ci - d\sqrt{2}i \tag{7.20b}$$

The product ST is another automorphism of K. Here is a list of the effects of these automorphisms on $\sqrt{2}$ and i:

$$S(\sqrt{2}, i) \to (-\sqrt{2}, i), \qquad T(\sqrt{2}, i) \to (\sqrt{2}, -i)$$
$$ST(\sqrt{2}, i) \to (-\sqrt{2}, -i), \qquad E(\sqrt{2}, i) \to (\sqrt{2}, i) \tag{7.20c}$$

We note the following

Theorem. Any automorphism of T of finite extension K over F maps each element u of K to a conjugate Tu of u over F. By this theorem, any other automorphism must map $\sqrt{2}$ into a conjugate $\pm\sqrt{2}$ and i into a conjugate $\pm i$. These are the four possibilities (E, S, T, ST) listed above. We conclude that the elements in this set are the only automorphisms of K over $\mathbb{Q}$. The multiplication table for this group follows from (7.20c) and is given by

$$E, S^2 = E, T^2 = E, ST = TS \tag{7.20d}$$

which may be identified with the four group C_{2v} [see (1.9)].

Definition. The *automorphism group* of a field K over a subfield F is the group of those automorphisms of K which leave every element of K invariant.

Theorem. Any automorphism T of a finite extension K of F maps each elements u of K to a conjugate Tu of u over F. That is, u and Tu satisfy the same irreducible equation over F.

Proof. Consider the monic irreducible polynomial

$$p(x) = x^n + a_{n-1}x^{n-1} + \ldots + a_0 \qquad (7.21a)$$

The automorphism T preserves rational relations [e.g., (7.20)] and leaves each a_i fixed. With $p(u) = 0$ we obtain

$$T(u^n + a_{n-1}u^{n-1} + \ldots + a_0) = (Tu)^n + a_{n-1}(Tu)^{n-1} + \ldots + a_0 = 0) \quad (7.21b)$$

This relation indicates that Tu is also a root of $p(u)$ so that Tu is a conjugate of u.

Definition. If $N = F(u_1, \ldots, u_n)$ is the root field of a polynomial

$$f(x) = (x - u_i)(x - u_2) \ldots (x - u_n) \qquad (7.21c)$$

then the automorphism group of N over F is known as the *Galois group* of the equation $f(x) = 0$, or as the *Galois group of the field N over F*. (Root fields were described in Section 7.3.)

Automorphisms and Permutations

Let N be the root field of $f(x)$ over F. In this event, the automorphism T produces a permutation λ of the distinct roots $(u_1, \ldots, u_k)$ of $f(x)$ so that

$$Tu_i = u_{1\lambda}; \quad Tu_2 = u_{2\lambda}, \ldots, Tu_k = u_{k\lambda} \qquad (7.22a)$$

However, every element w in the root field may be expressed as the polynomial $w = h(u_1, \ldots, u_k)$ with coefficients in F. As T leaves these coefficients fixed, (7.22a) implies

$$T[h(u_1, \ldots, u_k)] = h(Tu_1, \ldots, Tu_k) = h(u_{1\lambda}, \ldots, u_{k\lambda}) \qquad (7.22b)$$

It follows that the effect of T on w is determined by the effect of T on the roots of w. One may conclude that the isomorphism T is uniquely determined by the permutation (7.22a). The product of two such permutations is obtained by applying the related automorphisms in succession. It follows that the permutations (7.22a) form a group isomorphic to the group of automorphisms. These findings are summarized in the following:

Theorem. Let $f(x)$ be any polynomial of degree n over the field F which has k distinct roots $u_1, \ldots, u_k$ in a root field $N = F(u_1, \ldots, u_k)$. Then each automorphism T of the Galois group G of $f(x)$ determines a permutation $u_i \leftrightarrow Tu_i$ of the distinct roots of $f(x)$ and, conversely, the automorphism T is completely determined by this permutation.

The following corollaries apply:

1. The Galois group of any polynomial is isomorphic to a group of permutations of its distinct roots.

2. The order of a Galois group of a polynomial of degree n divides $n!$.

Symmetric Polynomial

The polynomial $g(x_1, \ldots, x_n)$ is symmetric iff it is invariant under the symmetric group of all permutations of its subscripts. Here are three examples:

$$g_1 = x_1 + x_2 + x_3, \quad g_2 = x_1x_2 + x_1x_3 + x_2x_3 \qquad (7.23)$$

$$g_3 = x_1x_2x_3$$

These symmetric polynomials are coefficients in the expansion

$$(t - x_1)(t - x_2)(t - x_3) = t^3 - g_1t^2 + g_2t^2 - g_3$$

With the preceding corollary we note that the Galois group of this polynomial is isomorphic to the symmetric group, S_3, on the three letters, $\{x_1, x_2, x_3\}$ (see Section 4.5).

Definition. A finite extension N over a field F is *normal* if every polynomial $p(x)$ irreducible over F that has one root in N has all its roots in N.

Theorem. A finite extension N is normal over F iff it is the root field of some polynomial over F.

Definition. A polynomial $f(x)$ of degree n is *separable* over a field F if it has n distinct roots in some root field $N \supset F$; otherwise, $f(x)$ is inseparable. A finite extension $K \supset F$ is called separable over F if every element in K satisfies a separable polynomial equation over F.

Definition. A polynomial equation $f(x) = 0$ with coefficients in F is *solvable by radicals* over F if its roots lie in an extension K of F obtainable by successive adjunction of nth roots.

Solvable Groups

Definition. A finite group G is called *solvable* iff it contains a chain ('tower') of subgroups,

$$G = Q_0 \supset Q_1 \supset Q_2 \supset \ldots \supset Q_s = E \qquad (7.24a)$$

such that for all k, Q_k is an invariant subgroup of Q_{k-1} and the factor group Q_{k-1}/Q_k is cyclic. It is known, for example, that any group whose order is divisible by less than three distinct primes is solvable. Every group of prime order is solvable. The sequence (7.24a) is labeled a *composition series* and the set of subgroups in the sequence *composition factors*. In this context we note the *Jordan-Hölder theorem* which states that any two composition series of a finite group are isomorphic.

Theorem. If a polynomial equation $f(x) = 0$ with coefficients in F is solvable by radicals, then its Galois group over F is solvable.

The concept of a polynomial equation being solvable by radicals (i.e., by the arithmetic operations of addition, subtraction, multiplication, division and the extraction of roots, is familiar with respect to the quadratic equation

$$x^2 + bx + c = 0 \qquad (7.24b)$$

whose solution is given by the well known expression

$$x = \frac{-b \pm \sqrt{b^2 - 4c}}{2} \qquad (7.24c)$$

The question Galois' theorem addresses is the following: What is the highest degree of a polynomial equation which, in general, may be solved by radicals? The analysis continues with the following analysis.

In Section 4.5, the elements of a permutation group were written in terms of cyclic components. A cyclic component with q terms is called a *q-cycle*. Thus, the symmetric group S_3 contains the identity, E, the 2-cycles (12), (13), (23) and the 3-cycles (123), (132). (The related group table is shown on page 73.) In this representation, a missing index is unchanged in the permutation, so that the cycle (13) means (13)(2). The *period* of a cycle is the number of changes required to return the cycle. For example, the 3-cycle (123)→ (231) → (312) = (123) has period 3. The period of a q-cycle is q. Every permutation can be written as a product of disjoint cycles which act independently. We recall that if G is a group, then G and E are invariant subgroups of G. These invariant subgroups are tacitly included or explicitly included in the statements of theorems to follow.

Definition. A *simple group* is a group that has no invariant subgroups other than itself and the identity. A permutation group that contains only even permutations is called the *alternating group*, and carries the symbol A_n (see also Section 4.5).

Lemma. The period of a permutation is the least common period of constituent cycles.

Thus the period of S_3 is two. Consider the 'cube' of a 3-cycle (in cyclic order):

$$(312)(231)(123) = (1)(2)(3) = E \qquad (7.25a)$$

It follows that if τ is a q-cycle, then

$$(\tau)^q = E \qquad (7.25b)$$

The following will prove useful in transformation calculations. If κ is a given cycle decomposition, then elements of the similarity transformation $\sigma\kappa\sigma^{-1}$ are obtained by performing the permutation σ on the separate cycles of κ. Consider the following operation on seven numbers:

$$\sigma = (123)(567)(4)$$

$$\kappa = (314)(25)(67) \qquad (7.26a)$$

$$\sigma\kappa\sigma^{-1} = (124)(36)(75)$$

Transposition

Another representation of permutations which is useful is that of transpositions (i.e., 2-cycles). Related decompositions occur in the following manner:

$$(a_1 a_2 \cdots a_m) = (a_1 a_m)(a_1 a_{m-1})(a_1 a_{m-2}) \cdots (a_1 a_2) \qquad (7.26b)$$

and

$$(a_1 a_m)(a_1 a_{m-1}) = (a_1 a_{m-1} a_m) \qquad (7.26c)$$

From the first of these relations, it follows that any permutation on m letters may be written as a product of $m - 1$ transpositions. Here is another useful rule. Namely, the inverse of a permutation is obtained simply by reflecting the permutation through its mid-point. Thus $(1234)^{-1} = (4321)$; $(27458)^{-1} = (85472)$.

The A_n Group

The alternating group consists of even permutations. A cycle in an odd number of elements is even, otherwise odd. Thus a 3-cycle is even and a two cycle is odd. The product of two 2-cycles is even [$o \times o = e$, recall (4.42)]. The A_3 group consists of E and two 3-cycles, and is of order three [i.e., C_3 (1.5b)]. The group A_4 consists of E, three products of 2-cycles and eight 3-cycles and is of order twelve. We note that A_n is an invariant subgroup of S_n. Namely,

$$\sigma^{-1} A_n \sigma = A_n, \quad \sigma \in S_n \qquad (7.26d)$$

As the period of a permutation is preserved in a similarity transformation, the evenness of the elements of A_n maintains in (7.26d). This invariance is consistent with the following: In the event that σ is odd, $o \times e = o$, $o \times o = e$, and the evenness of elements of A_n is preserved. Here we recall that the evenness or oddness of the inverse of a permutation is equal to that of the original permutation.

Proposition I

If $G \neq E$ is an invariant subgroup of A_n or S_n and $n \neq 4$, then G contains either a 2-cycle or a 3-cycle.

Proof. Let $\kappa \in G$ have a prime period, p, so that with (7.25b), $\kappa^p = E$. In general if the period of κ is not prime, then it has a prime divisor whose factors have prime periods. Thus, with the preceding lemma, the cycle of decomposition of κ only contains p-cycles. The following possibilities pertain:

(a) $\kappa = (12)$, a 2-cycle, for which no further argument is required.

(b) $\kappa = (12)(34)\cdots$, 2-cycle products.

(c) $\kappa = (123)$ a 3-cycle, for which no further argument is required.

(d) $\kappa = (123)(456)\cdots$, 3-cycle products.

(e) $\kappa = (12345$ or more$)\cdots$, p-cycle products, $p > 3$.

We must consider cases (b), (d) and (e). As G is assumed to be an invariant subgroup of A_n or S_n, it follows that if $\kappa \in G, \sigma \in A_n$ we may write [recall (5.10)],

$$\sigma \kappa \sigma^{-1} \in G \text{ and } \sigma \kappa \sigma^{-1} \kappa^{-1} \in G \qquad (7.26e)$$

Case (d). Consider that $\kappa = (123)(456)\cdots$. Choosing $\sigma = (1234)$ gives (all remaining terms do not change)

$$\sigma \kappa \sigma^{-1} = (1234)\,[(123)\,(456)]\,(4321) = (156)\,(234)$$

$$\sigma \kappa \sigma^{-1} \kappa^{-1} = (41)(25)(3)(6) = (41)(25)$$

So in fact, case (d) is included in case (b).

Case (e). We set $\kappa = (12345 \cdots p)\cdots$. Choosing $\sigma = (234)$ we obtain

$$\sigma \kappa \sigma^{-1} = (13425 \cdots p)\cdots$$

Furthermore,

$$\sigma \kappa \sigma^{-1} \kappa^{-1} = (1)(352)(4)(6)\cdots(p) = (352)$$

It follows that for this case, G contains a 3-cycle.

Case (b). For this set, $\kappa = (12)(34) \cdots$. Choosing $\sigma = (123)$ gives

$$\sigma \kappa \sigma^{-1} = (23)(14) \cdots$$

$$\sigma \kappa \sigma^{-1} \kappa^{-1} = (13)(24)$$

There remain two possibilities:

(i) $(n > 4)$ If G contains a product of 2-cycles, $\kappa = (12)(34)$, then with $\sigma = (125)$ we obtain

$$\sigma \kappa \sigma^{-1} = (25)(34)$$

$$\sigma \kappa \sigma^{-1} \kappa^{-1} = (25)(34)(43)(21) = (25)(21)(3)(4) = (152)(3)(4) = (152)$$

so that G contains a 3-cycle. Thus, for $n \neq 4$, invariant subgroups of A_n or S_n contain either 2- or 3-cycles which establishes Proposition I.

(ii) $(n = 4)$ Again set $\kappa = (12)(34)$. With $\sigma = (123)$ we obtain

$$\sigma \kappa \sigma^{-1} = (23)(14)$$

$$\sigma \kappa \sigma^{-1} \kappa^{-1} = (13)(24)$$

These three products of 2-cycle terms, together with the identity, comprise an invariant fourth-order subgroup, F_4 (see Problem 7.34).

Corollary. The symmetric group S_4 is solvable. Namely, S_4 contains the invariant subgroup A_4 which in turn contains F_4 as an invariant subgroup, etc., so that

$$G = S_4 \supset A_4 \supset F_4 \cdots \supset E \tag{7.27}$$

and (7.24a) is satisfied.

Theorem. If $n \neq 4$, the only invariant subgroup of S_n is A_n or S_n.

Proof. The cases $n = 2, 3$ are elementary (cf., S_3 group table on p. 73). Consider $n > 4$. With Proposition I, if G is an invariant subgroup it must contain either a 2-cycle or a 3-cycle. For a 2-cycle we choose $\sigma = (12)$ and $\kappa = (234)$. Then G contains all 2-cycles, namely,

$$(234)(12)(234)^{-1} = (13) \in G$$

We conclude that G is the whole symmetric group S_n since any element in G can be represented as a product of 2-cycle transpositions. In like manner, if G contains any 3-cycle, it contains all 3-cycles. In this context we demonstrate that any digit of a permutation can be transformed into any other, such as, for example, $(123) \rightarrow (124)$. Consider that $\sigma = (123)^2 = (132) \in G$. With $\kappa = (12)(34)$ we find

$$\kappa \sigma^2 \kappa^{-1} = (124)$$

We conclude that G contains every 3-cycle.

Proposition II

Every even permutation can be expressed as a product of 3-cycles. Namely, represent the permutation as a product of transpositions. As the given permutation is even (it contains an odd number of elements) the number of transpositions is likewise even. These in turn may be paired so that the first goes with the second, the third with the fourth, etc. There are two possibilities:

(a) The two transpositions have one element in common in which case the product gives a 3-cycle, namely [with (7.26b)],

$$(12)(13) = (132) \tag{7.28a}$$

(b) The two transpositions of a pair have no common digits. For this case we obtain

$$(12)(34) = (12)E(34) = (12)(13) \times (31)(34) = (132)(341), \quad (7.28b)$$

a product of 3-cycles, which establishes Proposition II.

As the implied inclusive relation between G and A_n was obtained from all relevant cycles, we conclude:

Theorem A. The only invariant subgroup of G is A_n. But G is identified with $\mathcal{S}_n$ so that A_n is the only invariant subgroup of $\mathcal{S}_n$. Note that this conclusion is consistent with (7.26d) which states that A_n is an invariant subgroup of $\mathcal{S}_n$.

With these results we are prepared to prove:

Theorem B. The symmetric group $\mathcal{S}_n$ on n letters $\{x_1, \cdots, x_n\}$ is not solvable for $n > 4$.

Proof. As the symmetric group for $n > 4$ contains only one invariant subgroup A_n, which is a simple group, it follows that the symmetric group on n letters, for $n > 4$, is not solvable. [Namely, (7.24a) is not satisfied.]

Theorem C. There exists a real quintic equation whose Galois group is the symmetric group on $\{x_1, x_2, x_3, x_4, x_5\}$.

Proof. Consider the field of all algebraic numbers $\mathbb{A}$ which is countable and contains all roots of unity. One may choose five algebraically independent real numbers $x_1, \ldots, x_5$ over $\mathbb{A}$ with which the transcendental extension $\mathbb{A}(x_1, \ldots, x_5)$ may be formed. Let $g_1, \ldots, g_5$ be elementary symmetric polynomials in x_i and let $F = \mathbb{A}(g_1, \ldots, g_5)$. With preceding theorems, we note that the Galois group of the polynomial over F,

$$f(t) = t^5 - g_1 t^4 + g_2 t^3 - g_3 t^2 + g_4 t - g_5 = 0 \qquad (7.29)$$

is the symmetric group on $\{x_1, x_2, x_3, x_4, x_5\}$.

Incorporating this result with *Theorem C* indicates that there exists a real quintic equation over a field containing all roots of unity whose Galois group is not solvable. We may therefore conclude the following:

There exists a real quintic equation which is not solvable by radicals. (This is Galois' famous result obtained before his tragic death at the age of 20.)[5]

---

[5]For further discussion, see E.T. Bell, *Men of Mathematics*, Dover, New York, (1937).

Symbols for Chapter 7

$\mathbb{A}$ = Set of algebraic numbers.

$\mathbb{C}$ = Set of complex numbers.

R = A ring.

$\mathbb{R}$ = Set of real numbers.

$\mathbb{Q}$ = Set of rationals.

$\mathbb{Q}(\sqrt{n}) = a + \sqrt{n}\,b$, (a, b) are rational numbers and $\sqrt{n}$ is a root of a polynomial. (A simple algebraic extension of the rationals, $\mathbb{Q}$.)

$\mathbb{Z}$ = Set of integers.

$\mathbb{Z}_n$ = Set of residue classes (mod n).

$F[x]$ = Domain of polynomial forms in the indeterminate x over the field F.

A polynomial and the permutation group have *degrees*. A group has an *order*. The permutation group Γ_n is of order n and degree $n!$

$[u{:}F]$ = The degree of a monic irreducible polynomial with coefficients in F and root u.

$F(y) = f(y)/g(y)$ = Field generated by F and y.

$a|b$ Indicates that a divides b.

$(a, b) = gcd$ of a, b.

Summary of Topics for Chapter 7

1. Integral domains, rings and fields.

2. Associative and commutative rings.

3. Isomorphisms, homomorphisms and automorphisms.

4. Ideal of a ring.

5. Algebraic numbers.

6. Congruences and modulo.

7. Algebra of congruences.

8. Greatest common divisor (gcd).

9. Chinese remainder theorem.

10. Irreducible polynomials.

11. Eisenstein's irreducibility criterion.

12. gcd of two polynomials.

13. Extension of a field.

14. Root fields.

15. Conjugate over a field.

16. Monic polynomial.

17. Elements algebraic over a field.

18. Automorphism group; Galois group.

19. Symmetric polynomials.

20. Normal extension.

21. Composition series and composition factors.

22. Solvable group.

23. Equation solvable by radicals.

24. Galois theorem.

Problems

7.1 Show that the set $\mathbb{Z}(\sqrt{m})$ is an integral domain.

Answer

Consider the closure property (i) for addition and multiplication, respectively,

$$(a + b\sqrt{m}) + (c + d\sqrt{m}) = (a + c) + (b + d)\sqrt{m}$$
$$(a + b\sqrt{m})(c + d\sqrt{m}) = (ac + mbd) + (ad + bc)\sqrt{m}$$

where a, b, c, d are integers. Terms on the right of each equation are elements of $\mathbb{Z}(\sqrt{m})$.

7.2 Show that an isomorphism between two integral domains preserves differences.

7.3 Show that the field of rational numbers, $\mathbb{Q}$, is a subfield of the extended field $\mathbb{Q}(\sqrt{2})$.

7.4 Consider the number $N = q^r$ where q is an integer and r is a rational number. What type of number is N? Verify your answer.

7.5 Show that $k^{1/n}$, where k is an irrational number and n is an integer, is irrational.

Answer

Set $k^{1/n} = q/s$ where $q \leq s$ are integers. Taking the nth power of both sides gives $k = q^n/s^n$, which is a contradiction.

7.6 Show that the set $\{0, 1\}$ with addition and multiplication defined as usual, except $1 + 1 = 0$ (instead of 2), is an integral domain. This set is, in fact, $\mathbb{Z}_2$.

7.7 Under what conditions is $F[x]$ a subfield?

7.8 Are the two statements, $a = 0 \pmod{p}$ and $a = 0 \pmod{p^2}$, where p is prime, consistent?

Answer

They are not. The first statement implies that one value of a is prime, viz., $a = kp$, $k = 1$. The second implies that no a is prime.

7.9 Which of the sets $\mathbb{Q}$, $\mathbb{R}$, $\mathbb{C}$, $\mathbb{Z}$ and $\mathbb{A}$ are fields? What are the corresponding operations for this property to be valid? Which of these fields are algebraically complete?

7.10 The symbol $\mathbb{Z}_n$ denotes the set of integers mod n. (a) What are positive integers of the set $n \pmod{5}$ where n is an integer? *Hint:* Recall (7.4b,c). (b) Show that the set $\mathbb{Z}_5$ is a finite integral domain. (c) Prove that $\mathbb{Z}_5$ is a finite field.

7.11 If $f(x)$ is irreducible over the field F, show that $f(x + a)$, for any $a \in F$, is likewise irreducible over the field F.

7.12 Over what fields, F, is the polynomial, $x^2 + 1$, irreducible?

7.13 What is the root field of the polynomial, $f(x) = x^3 - x^2 - 2x + 1$?

7.14 Test Fermat's theorem (7.3c) for $p = 3$ and $a = 11$.

7.15 For what integers, n, is the gcd, $(n, p) = 1$, where p is a prime?

7.16 Which polynomials are relatively prime to $x^3 + 3x^2 + 1$?

7.17 Is a commutative ring with an identity element an integral domain?
Answer
No. The ring must also include the implication postulate ($D5$).

7.18 The *basis* of an extended field are terms which, when multiplied by elements of the original field, generate the extended field. For example, consider the extension of the rational field $\mathbb{Q}(\sqrt{2})$: $1a + \sqrt{2}\,b$. This form indicates that $(1, \sqrt{2})$ forms a basis for the field $\mathbb{Q}(\sqrt{2})$. That is, elements of the field are obtained by multiplying elements of the basis by rational numbers. What are the basis element of the field $\mathbb{Q}(\sqrt{2}, i)$?
Answer
Elements of this extended field are given by

$$w = (a + \sqrt{2}\,b) + (c + d\sqrt{2})i = a + \sqrt{2}\,b + ci + d\sqrt{2}\,i$$

where a, b, c, d are rational numbers. It follows that this field has the basis $(1, \sqrt{2}, i)$.

7.19 Show that the field $\mathbb{Q}(\sqrt{2}, i)$ is the root field of the polynomial, $x^4 - 2x^2 + 9$, thereby establishing that this field is of degree 4 over the field of rationals.

7.20 Show that the set consisting only of 0 with $0 + 0 = 0 \cdot 0 = 0$, satisfies all postulates of an integral domain except $0 \neq 1$ in (f).

7.21 Is the set of integers, $\mathbb{Z}$, a field?
Answer
As the elements of this set do not include inverses, it is not a field.

7.22 (a) Consider an nth degree monic polynomial, which has n real roots. What is the factored form of this polynomial? (b) Consider an nth monic polynomial with integer coefficients whose constant term is a prime number. If all the roots of this polynomial are real and one of the roots is a prime number, what are the remaining $n - 1$ roots?
Answers
(a)

$$p(x) = (x - k_1)(x - k_2)\ldots(x - k_n)$$

where k_i is the ith root.

(b) From (a) and the fact that the constant term is a prime number, we see that the remaining $n-1$ roots are each 1.

7.23 Prove that any finite Abelian group is solvable.

7.24 Prove that the permutation group on three objects is solvable.

7.25 Prove that if a group, G, has an invariant subgroup, B, such that B and the factor group G/B are both solvable, then G is solvable.

7.26 Consider that the ring R has an identity $1 \neq 0$. An element $u \in R$ is called a *unit* in R if there is some $v \in R$ such that $uv = vu = 1$. (a) Show that any element $a \neq 0$ of a field is a unit. (b) Show that the units of the ring Z of integers are ± 1. (c) Consider the numbers $a + b\sqrt{2} \in Z[\sqrt{2}]$. Show that the units of $Z[\sqrt{2}]$ are $1 \pm \sqrt{2}$ and $3 \pm 2\sqrt{2}$.

Hint [for (c)]: Units are numbers $(x + y\sqrt{2})$ which are solutions of

$$(a + b\sqrt{2})(x + y\sqrt{2}) = 1$$

where x, y are integers.

7.27 (a) Show that every positive integer has a prime factor. (b) Prove *the fundamental theorem of arithmetic*: Every positive integer may be written as a product of prime numbers. (c) Show that this product is unique.

Answers (partial)

(a) If the integer n is non-prime, there exists integers a_1, a_2 such that $n = a_1 a_2$. Then $a_1 < n$ and a_2 is either prime or another composite. If a_1 is prime the theorem is complete. If not, then it too is a composite and we may write $a_1 = a_3 a_4$, where $0 < a_3 < a_1$. If a_3 is not a prime, we may continue in this manner, but only for a finite number of steps as

$$n > a_1 > a_3 > \ldots > 0$$

In this manner we arrive at a factor a_n that is divisible only by itself and unity and is therefore a prime which proves the theorem.

(b) Consider the positive integer n. By the preceding theorem n has a prime factor, p_1, so that $n = p_1 n_1$. If n_1 is not prime we again apply this theorem and write $n_1 = p_2 n_2$ so that $n = p_1 p_2 n_2$. As $n > n_1 > n_2 > \ldots > 0$, it suffices to carry out this process a finite number of times until n is factored into positive primes.

7.28 Show that

$$2p \mid [(p-1)! - p + 1] \Rightarrow p \mid [(p-1)! + 1]$$

Answer

From the given information,

$$(p-1)! - p + 1 = k2p$$

where k is a positive integer. Thus,

$$(p-1)! + 1 = k^3 p$$

and we may conclude that

$$p | [(p-1)! + 1]$$

7.29 Show that the nth root of a positive integer is either an integer or an irrational number. *Hint*: Employ the fundamental theorem of arithmetic and assume that the given statement is not true.

7.30 Let n be an integer and let p be a prime such that $n \geq p$. Let $n \pmod{p}$ denote the minimum of the absolute value of this form. For example, $7 \pmod{3} = 1, 10 \pmod{4} = 2$, etc. Show that (a) $n \pmod{p}$ is the remainder of n/p and (b) at fixed p, $n \pmod{p}$ is a group under addition. That is,

$$n \pmod{p} + n' \pmod{p} = (n + n') \pmod{p}$$

(c) This group is of order p. [The 'mod p group.']

Proof (partial): In (c), you are asked to show that $n \pmod{p}$ comprises p distinct values. In this scheme, $n \pmod{p} = 0$ corresponds to $n = kp$, where k is an integer; 1 corresponds to $n = kp + 1$; and $p - 1$ corresponds to $n = p(k+1) - 1$. As this series increases by unit steps, there are $p - 1$ values in all. For example, in calculating $n \pmod{3}$ we find $4 \pmod{3} = 1, 5 \pmod{3} = 2, 6 \pmod{3} = 0, 7 \pmod{3} = 1$, etc., which give the three values $(0, 1, 2)$. In addition, say, $4 \pmod{3} + 7 \pmod{3} = 1 + 1 = 11 \pmod{3} = 2$.

7.31 Show that the square root of an even number that is not a perfect square is irrational.
Answer
First we introduce the

Lemma. If q^2 is odd/even, then q is odd/even.

If the $\sqrt{2}$ is rational we may set

$$p^2 = 2q^2 \tag{P1}$$

where either p is even and q is odd or vice versa. From (P1) it follows that $2 | p^2$ so that with the above lemma, p is even. Then we may set $p \equiv 2z$ so that $2z^2 = q^2$. With the stated lemma this implies that q is also even which is a contradiction with our starting hypothesis. We may conclude that $\sqrt{2}$ is irrational.

Now let x be an even number that is not a perfect square. Then we may write

$$x = 2z \tag{P2}$$

where z is an integer such that $2\!\!\not|z$. With the fundamental theorem of arithmetic, z is a product of primes. Let

$$\sqrt{x} = \sqrt{2}\sqrt{z} \tag{P3}$$

It follows that $\sqrt{x}$ irrational.

7.32 (a) For what order, n, is the symmetric group, S_n, not solvable? (b) State a condition on the order n of a group which implies that the group is solvable.

7.33 Determine explicitly whether or not the groups S_2, S_3 are solvable, respectively.

7.34 Consider the 2-cycle products listed in case (b(ii)), page 183. These components, together with the identity, comprise the fourth-order group, F_4. Identify this group [see Section (1.1)].

7.35 (a) What roots of unity are the solutions of the following equation:

$$x^2 + x + 1 = 0?$$

(b) What class of equations does this equation belong to?

7.36 What are the invariant subgroups of S_n?
Answer
The permutation group on n letters has only one invariant subgroup, the symmetric group A_n.

Appendix A. Character Tables for the Point Groups[1]

The C_1, C_s, and C_i Groups

C_1	E
A	1

C_s	E	σ_h		
A'	1	1	x, y, R_z	$x^2, y^2,$ z^2, xy
A''	1	-1	z, R_x, R_y	yz, xz

C_i	E	i		
A_g	1	1	R_x, R_y, R_z	x^2, y^2, z^2 xy, xz, yz
A_u	1	-1	x, y, z	

[1]From: F.A. Cotton, *Chemical Applications of Group Theory*, 3rd ed., Copyright (1963), John Wiley. Reprinted by permission of John Wiley & Sons, Inc.

The C_n Groups[2]

C_2	E	C_2		
A	1	1	z, R_z	x^2, y^2, z^2, xy
B	1	-1	x, y, R_x, R_y	yz, xz

C_3	E	C_3	C_3^2		$\varepsilon = \exp(2\pi i/3)$
A	1	1	1	z, R_z	$x^2 + y^2, z^2$
E	1	ε	ε^*		$(x^2 - y^2, xy)(yz, xz)$
	1	ε^*	ε	$(x, y)(R_x, R_y)$	

C_4	E	C_4	C_2	C_4^3		
A	1	1	1	1	z, R_z	$x^2 + y^2, z^2$
B	1	-1	1	-1		$x^2 - y^2, xy$
E	1	i	-1	$-i$		
	1	$-i$	-1	i	$(x, y)(R_z, R_y)$	(yz, xz)

C_5	E	C_5	C_5^2	C_5^3	C_5^4		$\varepsilon = \exp(2\pi i/5)$
A	1	1	1	1	1	z, R_z	$x^2 + y^2, z^2$
E_i	1	ε	ε^2	ε^{2*}	ε^*	$(x, y)(R_x, R_y)$	(yz, xz)
	1	ε^*	ε^{2*}	ε^2	ε		
E_2	1	ε^2	ε^*	ε	ε^{2*}		$(x^2 - y^2, xy)$
	1	ε^{2*}	ε	ε^*	ε^2		

---

[2]Note that the E label of the first column of a character table represents the identity element. The parameter E under the group-title symbol (e.g., C_4), represents a two-dimensional irrep of the group. The symbols A, B each represent one-dimensional irreps, T, three-dimensional irreps, G, four-dimensional irreps, and H, five-dimensional irreps. In the event that the two-dimensional E irrep partitions into two one-dimensional irreps, related 1-values in the table are slightly separated.

C_6	E	C_6	C_3	C_2	C_3^2	C_6^5		$\varepsilon = \exp(2\pi i/6)$
A	1	1	1	1	1	1	z, R_z	x^2+y^2, z^2
B	1	-1	1	-1	1	-1		
	1	ε	$-\varepsilon^*$	-1	$-\varepsilon$	ε^*		(yz, yz)
E_1							(x,y)	
							(R_x, R_y)	
	1	ε^*	$-\varepsilon$	-1	$-\varepsilon^*$	ε		
	1	$-\varepsilon^*$	$-\varepsilon$	1	$-\varepsilon^*$	$-\varepsilon$		(x^2-y^2, xy)
E_2								
	1	$-\varepsilon$	$-\varepsilon^*$	1	$-\varepsilon$	$-\varepsilon^*$		

C_7	E	C_7	C_7^2	C_7^3	C_7^4	C_7^5	C_7^6		$\varepsilon = \exp(2\pi i/7)$
A	1	1	1	1	1	1	1	z, R_z	x^2+y^2, z^2
	1	ε	ε^2	ε^3	ε^{3*}	ε^{2*}	ε^*		
E_1								(x,y)	(xz, yz)
								(R_x, R_y)	
	1	ε^*	ε^{2*}	ε^{3*}	ε^3	ε^2	ε		
	1	ε^2	ε^{3*}	ε^*	ε	ε^3	ε^{2*}		
E_2									(x^2-y^2, xy)
	1	ε^{2*}	ε^3	ε	ε^*	ε^{3*}	ε^2		
	1	ε^{3*}	ε^*	ε^2	ε^{2*}	ε	ε^{3*}		
E_3									
	1	ε^{3*}	ε	ε^{2*}	ε^2	ε^*	ε^3		

C_8	E	C_8	C_4	C_8^3	C_2	C_8^5	C_4^3	C_8^7		$\varepsilon = \exp(2\pi i/8)$
A	1	1	1	1	1	1	1	1	z, R_z	x^2+y^2, z^2
B	1	-1	1	-1	1	-1	1	-1		
E_1	1	ε	i	$-\varepsilon^*$	-1	$-\varepsilon$	$-i$	ε^*	(x,y)	(xz, yz)
	1	ε^*	$-i$	$-\varepsilon$	-1	$-\varepsilon^*$	i	ε	(R_x, R_y)	
E_2	1	i	-1	$-i$	1	i	-1	$-i$		(x^2-y^2, xy)
	1	$-i$	-1	i	1	$-i$	-1	i		
E_3	1	$-\varepsilon$	i	ε^*	-1	ε	$-i$	$-\varepsilon^*$		
	1	$-\varepsilon^*$	$-i$	ε	-1	ε^*	i	$-\varepsilon$		

The D_n Groups

D_2	E	$C_2(z)$	$C_2(y)$	$C_2(x)$		
A	1	1	1	1		x^2, y^2, z^2
B_1	1	1	-1	-1	z, R_z	xy
B_2	1	-1	1	-1	y, R_y	xz
B_3	1	-1	-1	1	x, R_x	yz

D_3	E	$2C_3$	$3C_2$		
A_1	1	1	1		x^2, y^2, z^2
A_2	1	1	-1	z, R_z	
E	2	-1	0	$(x,y)(R_x, R_y)$	$(x^2 - y^2, xy)(xz, yz)$

D_4	E	$2C_4$	$C_2 (= C_4^2)$	$2C_2'$	$2C_2''$		
A_1	1	1	1	1	1		$x^2 + y^2, z^2$
A_2	1	1	1	-1	-1	z, R_z	
B_1	1	-1	1	1	-1		$x^2 - y^2$
B_2	1	-1	1	-1	1		xy
E	2	0	-2	0	0	$(x,y)(R_x, R_y)$	(xz, yz)

D_5	E	$2C_5$	$2C_5^2$	$5C_2$		
A_1	1	1	1	1		x^2, y^2, z^2
A_2	1	1	1	-1	z, R_z	
E_1	2	$2\cos 72°$	$2\cos 144°$	0	$(x,y)(R_x, R_y)$	(xz, yz)
E_2	2	$2\cos 144°$	$2\cos 72°$	0		$(x^2 - y^2, xy)$

D_6	E	$2C_6$	$2C_3$	C_2	$3C_2'$	$3C_2''$		
A_1	1	1	1	1	1	1		$x^2 + y^2, z^2$
A_2	1	1	1	1	-1	-1	z, R_z	
B_1	1	-1	1	-1	1	-1		
B_2	1	-1	1	-1	-1	1		
E_1	2	1	-1	-2	0	0	$(x,y)(R_x, R_y)$	(xz, yz)
E_2	2	-1	-1	2	0	0		$(x^2 - y^2, xy)$

The C$_{nv}$ Groups

C_{2v}	E	C_2	$\sigma_v(xz)$	$\sigma_v'(yz)$		
A_1	1	1	1	1	z	x^2, y^2, z^2
A_2	1	1	-1	-1	R_z	xy
B_1	1	-1	1	-1	x, R_y	xz
B_2	1	-1	-1	1	y, R_x	yz

C_{3v}	E	$2C_3$	$3\sigma_v$		
A_1	1	1	1	z	x^2-y^2, z^2
A_2	1	1	-1	R_z	
E	2	-1	0	$(x,y)(R_x, R_y)$	$(x^2-y^2, xy)(xz, yz)$

C_{4v}	E	$2C_4$	C_2	$2\sigma_v$	$2\sigma_d$		
A_1	1	1	1	1	1	z	x^2+y^2, z^2
A_2	1	1	1	-1	-1	R_z	
B_1	1	-1	1	1	-1		x^2-y^2
B_2	1	-1	1	-1	1		xy
E	2	0	-2	0	0	$(x,y)(R_x, R_y)$	(xz, yz)

C_{5v}	E	$2C_5$	$2C_5^2$	$5\sigma_v$		
A_1	1	1	1	1	z	x^2+y^2, z^2
A_2	1	1	1	-1	R_z	
E_1	2	$2\cos 72°$	$2\cos 144°$	0	$(x,y)(R_x, R_y)$	(xz, yz)
E_2	2	$2\cos 144°$	$2\cos 72°$	0		(x^2-y^2, xy)

C_{6v}	E	$2C_6$	$2C_3$	C_2	$3\sigma_v$	$3\sigma_d$		
A_1	1	1	1	1	1	1	z	x^2+y^2, z^2
A_2	1	1	1	1	-1	-1	R_z	
B_1	1	-1	1	-1	1	-1		
B_2	1	-1	1	-1	-1	1		
E_1	2	1	-1	-2	0	0	$(x,y)(R_x, R_y)$	(xz, yz)
E_2	2	-1	-1	2	0	0		(x^2-y^2, xy)

The C$_{nh}$ Groups

C_{2h}	E	C_2	i	σ_h		
A_g	1	1	1	1	R_z	x^2, y^2, z^2, xy
B_g	1	-1	1	-1	R_x, R_y	xz, yz
A_u	1	1	-1	-1	z	
B_u	1	-1	-1	1	x, y	

C_{3h}	E	C_3	C_3^2	σ_h	S_3	S_3^5		$\varepsilon = \exp(2\pi i/3)$
A'	1	1	1	1	1	1	R_z	$x^2 + y^2, z^2$
E'	1	ε	ε^*	1	ε	ε^*	(x, y)	$(x^2 - y^2, xy)$
	1	ε^*	ε	1	ε^*	ε		
A''	1	1	1	-1	-1	-1	z	
E''	1	ε	ε^*	-1	$-\varepsilon$	$-\varepsilon^*$	(R_x, R_y)	(xz, yz)
	1	ε^*	ε	-1	$-\varepsilon^*$	$-\varepsilon$		

C_{4h}	E	C_4	C_2	C_4^3	i	S_4^3	σ_h	S_4		
A_g	1	1	1	1	1	1	1	1	R_z	$x^2 + y^2, z^2$
B_g	1	-1	1	-1	1	-1	1	-1		$x^2 - y^2, xy$
E_g	1	i	-1	$-i$	1	i	-1	$-i$	(R_x, R_y)	(xz, yz)
	1	$-i$	-1	i	1	$-i$	-1	i		
A_u	1	1	1	1	-1	-1	-1	-1	z	
B_u	1	-1	1	-1	-1	1	-1	1		
E_u	1	i	-1	$-i$	-1	$-i$	1	i	(x, y)	
	1	$-i$	-1	i	-1	i	1	$-i$		

C_{5h}	E	C_5	C_5^2	C_5^3	C_5^4	σ_h	S_5	S_5^7	S_5^3	S_5^9		$e=\exp(2\pi i/5)$
A'	1	1	1	1	1	1	1	1	1	1	R_z	x^2+y^2, z^2
E_1'	1	ϵ	ϵ^2	ϵ^{2*}	ϵ^*	1	ϵ	ϵ^2	ϵ^{2*}	ϵ^*	(x,y)	
	1	ϵ^*	ϵ^{2*}	ϵ^2	ϵ	1	ϵ^*	ϵ^{2*}	ϵ^2	ϵ		
E_2'	1	ϵ^2	ϵ^*	ϵ	ϵ^{2*}	1	ϵ^2	ϵ^*	ϵ	ϵ^{2*}		(x^2-y^2, xy)
	1	ϵ^{2*}	ϵ	ϵ^*	ϵ^2	1	ϵ^{2*}	ϵ	ϵ^*	ϵ^2		
A''	1	1	1	1	1	-1	-1	-1	-1	-1	z	
E_1''	1	ϵ	ϵ^2	ϵ^{2*}	ϵ^*	-1	$-\epsilon$	$-\epsilon^2$	$-\epsilon^{2*}$	$-\epsilon^*$	(R_x, R_y)	(xz, yz)
	1	ϵ^*	ϵ^{2*}	ϵ^2	ϵ	-1	$-\epsilon^*$	$-\epsilon^{2*}$	$-\epsilon^2$	$-\epsilon$		
E_2''	1	ϵ^2	ϵ^*	ϵ	ϵ^{2*}	-1	$-\epsilon^2$	$-\epsilon^*$	$-\epsilon$	$-\epsilon^{2*}$		
	1	ϵ^{2*}	ϵ	ϵ^*	ϵ^2	-1	$-\epsilon^{2*}$	$-\epsilon$	$-\epsilon^*$	$-\epsilon^2$		

C_{6h}	E	C_6	C_3	C_2	C_3^2	C_6^5	i	S_3^5	S_6^5	σ_h	S_6	S_3		$e=\exp(2\pi i/6)$
A_g	1	1	1	1	1	1	1	1	1	1	1	1	R_z	x^2+y^2, z^2
B_g	1	-1	1	-1	1	-1	1	-1	1	-1	1	-1		
E_{1g}	1	ϵ	$-\epsilon^*$	-1	$-\epsilon$	ϵ^*	1	ϵ	$-\epsilon^*$	-1	$-\epsilon$	ϵ^*	(R_x, R_y)	(xz, yz)
	1	ϵ^*	$-\epsilon$	-1	$-\epsilon^*$	ϵ	1	ϵ^*	$-\epsilon$	-1	$-\epsilon^*$	ϵ		
E_{2g}	1	$-\epsilon^*$	$-\epsilon$	1	$-\epsilon^*$	$-\epsilon$	1	$-\epsilon^*$	$-\epsilon$	1	$-\epsilon^*$	$-\epsilon$		(x^2-y^2, xy)
	1	$-\epsilon$	$-\epsilon^*$	1	$-\epsilon$	$-\epsilon^*$	1	$-\epsilon$	$-\epsilon^*$	1	$-\epsilon$	$-\epsilon^*$		
A_u	1	1	1	1	1	1	-1	-1	-1	-1	-1	-1	z	
B_u	1	-1	1	-1	1	-1	-1	1	-1	1	-1	1		
E_{1u}	1	ϵ	$-\epsilon^*$	-1	$-\epsilon$	ϵ^*	-1	$-\epsilon$	ϵ^*	1	ϵ	$-\epsilon^*$	(x,y)	
	1	ϵ^*	$-\epsilon$	-1	$-\epsilon^*$	ϵ	-1	$-\epsilon^*$	ϵ	1	ϵ^*	$-\epsilon$		
E_{2u}	1	$-\epsilon^*$	$-\epsilon$	1	$-\epsilon^*$	$-\epsilon$	-1	ϵ^*	ϵ	-1	ϵ^*	ϵ		
	1	$-\epsilon$	$-\epsilon^*$	1	$-\epsilon$	$-\epsilon^*$	-1	ϵ	ϵ^*	-1	ϵ	ϵ^*		

The D$_{nh}$ Groups

D_{2h}	E	$C_2(z)$	$C_2(y)$	$C_2(x)$	i	$\sigma(xy)$	$\sigma(xz)$	$\sigma(yz)$		
A_g	1	1	1	1	1	1	1	1		x^2+y^2, z^2
B_{1g}	1	1	-1	-1	1	1	-1	-1	R_z	xy
B_{2g}	1	-1	1	-1	1	-1	1	-1	R_y	xz
B_{3g}	1	-1	-1	1	1	-1	-1	1	R_x	yz
A_u	1	1	1	1	-1	-1	-1	-1		
B_{1u}	1	1	-1	-1	-1	-1	1	1	z	
B_{2u}	1	-1	1	-1	-1	1	-1	1	y	
B_{3u}	1	-1	-1	1	-1	1	1	-1	x	

D_{3h}	E	$2C_3$	$3C_2$	σ_h	$2S_3$	$3\sigma_v$		
A_1'	1	1	1	1	1	1		x^2+y^2, z^2
A_2'	1	1	-1	1	1	-1	R_z	
E'	2	-1	0	2	-1	0	(x,y)	(x^2-y^2, xy)
A_1''	1	1	1	-1	-1	-1		
A_2''	1	1	-1	-1	-1	1	z	
E''	2	-1	0	-2	1	0	(R_x, R_y)	(xz, yz)

D_{4h}	E	$2C_4$	C_2	$2C_2'$	$2C_2''$	i	$2S_4$	σ_h	$2\sigma_v$	$2\sigma_d$		
A_{1g}	1	1	1	1	1	1	1	1	1	1		x^2+y^2, z^2
A_{2g}	1	1	1	-1	-1	1	1	1	-1	-1	R_z	
B_{1g}	1	-1	1	1	-1	1	-1	1	1	-1		x^2-y^2
B_{2g}	1	-1	1	-1	1	1	-1	1	-1	1		xy
E_g	2	0	-2	0	0	2	0	-2	0	0	(R_x, R_y)	(xz, yz)
A_{1u}	1	1	1	1	1	-1	-1	-1	-1	-1		
A_{2u}	1	1	1	-1	-1	-1	-1	-1	1	1	z	
B_{1u}	1	-1	1	1	-1	-1	1	-1	-1	1		
B_{2u}	1	-1	1	-1	1	-1	1	-1	1	-1		
E_u	2	0	-2	0	0	-2	0	2	0	0	(x, y)	

D_{5h}	E	$2C_5$	$2C_5^2$	$5C_2$	σ_h	$2S_5$	$2S_5^3$	$5\sigma_v$		
A_1'	1	1	1	1	1	1	1	1		x^2+y^2, z^2
A_2'	1	1	1	-1	1	1	1	-1	R_z	
E_1'	2	$2\cos 72°$	$2\cos 144°$	0	2	$2\cos 72°$	$2\cos 144°$	0	(x, y)	
E_2'	2	$2\cos 144°$	$2\cos 72°$	0	2	$2\cos 144°$	$2\cos 72°$	0		(x^2-y^2, xy)
A_1''	1	1	1	1	-1	-1	-1	-1		
A_2''	1	1	1	-1	-1	-1	-1	1	z	
E_1''	2	$2\cos 72°$	$2\cos 144°$	0	-2	$-2\cos 72°$	$-2\cos 144°$	0	(R_x, R_y)	(xz, yz)
E_2''	2	$2\cos 144°$	$2\cos 72°$	0	-2	$-2\cos 144°$	$-2\cos 72°$	0		

D_{6h}	E	$2C_6$	$2C_3$	C_2	$3C_2'$	$3C_2''$	i	$2S_3$	$2S_6$	σ_h	$3\sigma_d$	$3\sigma_v$		
A_{1g}	1	1	1	1	1	1	1	1	1	1	1	1		x^2+y^2, z^2
A_{2g}	1	1	1	1	-1	-1	1	1	1	1	-1	-1	R_z	
B_{1g}	1	-1	1	-1	1	-1	1	-1	1	-1	1	-1		
B_{2g}	1	-1	1	-1	-1	1	1	-1	1	-1	-1	1		
E_{1g}	2	1	-1	-2	0	0	2	1	-1	-2	0	0	(R_x, R_y)	(xz, yz)
E_{2g}	2	-1	-1	2	0	0	2	-1	-1	2	0	0		(x^2-y^2, xy)
A_{1u}	1	1	1	1	1	1	-1	-1	-1	-1	-1	-1		
A_{2u}	1	1	1	1	-1	-1	-1	-1	-1	-1	1	1	z	
B_{1u}	1	-1	1	-1	1	-1	-1	1	-1	1	-1	1		
B_{2u}	1	-1	1	-1	-1	1	-1	1	-1	1	1	-1		
E_{1u}	2	1	-1	-2	0	0	-2	-1	1	2	0	0	(x, y)	
E_{2u}	2	-1	-1	2	0	0	-2	1	1	-2	0	0		

D_{8h}	E	$2C_8$	$2C_8^3$	$2C_4$	C_2	$4C_2'$	$4C_2''$	i	$2S_8$	$2S_8^3$	$2S_4$	σ_h	$4\sigma_d$	$4\sigma_v$		
A_{1g}	1	1	1	1	1	1	1	1	1	1	1	1	1	1		x^2+y^2, z^2
A_{2g}	1	1	1	1	1	-1	-1	1	1	1	1	1	-1	-1	R_z	
B_{1g}	1	-1	-1	1	1	1	-1	1	-1	-1	1	1	1	-1		
B_{2g}	1	-1	-1	1	1	-1	1	1	-1	-1	1	1	-1	1		
E_{1g}	2	$\sqrt{2}$	$-\sqrt{2}$	0	-2	0	0	2	$\sqrt{2}$	$-\sqrt{2}$	0	-2	0	0	(R_x, R_y)	(xz, yz)
E_{2g}	2	0	0	-2	2	0	0	2	0	0	-2	2	0	0		(x^2-y^2, xy)
E_{3g}	2	$-\sqrt{2}$	$\sqrt{2}$	0	-2	0	0	2	$-\sqrt{2}$	$\sqrt{2}$	0	-2	0	0		
A_{1u}	1	1	1	1	1	1	1	-1	-1	-1	-1	-1	-1	-1		
A_{2u}	1	1	1	1	1	-1	-1	-1	-1	-1	-1	-1	1	1	z	
B_{1u}	1	-1	-1	1	1	1	-1	-1	1	1	-1	-1	-1	1		
B_{2u}	1	-1	-1	1	1	-1	1	-1	1	1	-1	-1	1	-1		
E_{1u}	2	$\sqrt{2}$	$-\sqrt{2}$	0	-2	0	0	-2	$-\sqrt{2}$	$\sqrt{2}$	0	2	0	0	(x, y)	
E_{2u}	2	0	0	-2	2	0	0	-2	0	0	2	-2	0	0		
E_{3u}	2	$-\sqrt{2}$	$\sqrt{2}$	0	-2	0	0	-2	$\sqrt{2}$	$-\sqrt{2}$	0	2	0	0		

The D$_{nd}$ Groups

D_{2d}	E	$2S_4$	C_2	$2C_2'$	$2\sigma_d$		
A_1	1	1	1	1	1		x^2+y^2, z^2
A_2	1	1	1	-1	-1	R_z	
B_1	1	-1	1	1	-1		x^2-y^2
B_2	1	-1	1	-1	1	z	xy
E	2	0	-2	0	0	(x,y); (R_x, R_y)	(xz, yz)

D_{3d}	E	$2C_3$	$3C_2$	i	$2S_6$	$3\sigma_d$		
A_{1g}	1	1	1	1	1	1		x^2+y^2, z^2
A_{2g}	1	1	-1	1	1	-1	R_z	
E_g	2	-1	0	2	-1	0	(R_x, R_y)	(x^2-y^2, xy), (xz, yz)
A_{1u}	1	1	1	-1	-1	-1		
A_{2u}	1	1	-1	-1	-1	1	z	
E_u	2	-1	0	-2	1	0	(x,y)	

D_{4d}	E	$2S_8$	$2C_4$	$2S_8^3$	C_2	$4C_2'$	$4\sigma_d$		
A_1	1	1	1	1	1	1	1		x^2+y^2, z^2
A_2	1	1	1	1	1	-1	-1	R_z	
B_1	1	-1	1	-1	1	1	-1		
B_2	1	-1	1	-1	1	-1	1	z	
E_1	2	$\sqrt{2}$	0	$-\sqrt{2}$	-2	0	0	(x,y)	
E_2	2	0	-2	0	2	0	0		(x^2-y^2, xy)
E_3	2	$-\sqrt{2}$	0	$\sqrt{2}$	-2	0	0	(R_x, R_y)	(xz, yz)

D_{5d}	E	$2C_5$	$2C_5^2$	$5C_2$	i	$2S_{10}^3$	$2S_{10}$	$5\sigma_d$		
A_{1g}	1	1	1	1	1	1	1	1		x^2+y^2,z^2
A_{2g}	1	1	1	-1	1	1	1	-1	R_z	
E_{1g}	2	$2\cos 72°$	$2\cos 144°$	0	2	$2\cos 144°$	$2\cos 72°$	0	(R_x,R_y)	(xz,yz)
E_{2g}	2	$2\cos 144°$	$2\cos 72°$	0	2	$2\cos 72°$	$2\cos 144°$	0		(x^2-y^2,xy)
A_{1u}	1	1	1	1	-1	-1	-1	-1		
A_{2u}	1	1	1	-1	-1	-1	-1	1	z	
E_{1u}	2	$2\cos 72°$	$2\cos 144°$	0	-2	$-2\cos 144°$	$-2\cos 72°$	0	(x,y)	
E_{2u}	2	$2\cos 144°$	$2\cos 72°$	0	-2	$-2\cos 72°$	$-2\cos 144°$	0		

D_{6d}	E	$2S_{12}$	$2C_6$	$2S_4$	$2C_3$	$2S_{12}^5$	C_2	$6C_2'$	$6\sigma_d$		
A_1	1	1	1	1	1	1	1	1	1		x^2+y^2,z^2
A_2	1	1	1	1	1	1	1	-1	-1	R_z	
B_1	1	-1	1	-1	1	-1	1	1	-1		
B_2	1	-1	1	-1	1	-1	1	-1	1	z	
E_1	2	$\sqrt{3}$	1	0	-1	$-\sqrt{3}$	-2	0	0	(x,y)	
E_2	2	1	-1	-2	-1	1	2	0	0		(x^2-y^2,xy)
E_3	2	0	-2	0	2	0	-2	0	0		
E_4	2	-1	-1	2	-1	-1	2	0	0		
E_5	2	$-\sqrt{3}$	1	0	-1	$\sqrt{3}$	-2	0	0	(R_x,R_y)	(xz,yz)

The Sₙ Groups

S_4	E	S_4	C_2	S_4^3		
A	1	1	1	1	R_z	$x^2 + y^2, z^2$
B	1	-1	1	-1	z	$x^2 - y^2, xy$
E	1	i	-1	$-i$		
	1	$-i$	-1	i	$(y, y); (R_x, R_y)$	(xz, yz)

S_6	E	C_3	C_3^2	i	S_6^5	S_6		$\varepsilon = \exp(2\pi i/3)$
A_g	1	1	1	1	1	1	R_z	$x^2 + y^2, z^2$
E_g	1	ε	ε^*	1	ε	ε^*		$(x^2 - y^2, xy);$
	1	ε^*	ε	1	ε^*	ε	(R_x, R_y)	(xz, yz)
A_u	1	1	1	-1	-1	-1	z	
E_u	1	ε	ε^*	-1	$-\varepsilon$	$-\varepsilon^*$		
	1	ε^*	ε	-1	$-\varepsilon^*$	$-\varepsilon$	(x, y)	

S_8	E	S_8	C_4	S_8^3	C_2	S_8^5	C_4^3	S_8^7		$\varepsilon = \exp(2\pi i/8)$
A	1	1	1	1	1	1	1	1	R_z	$x^2 + y^2, z^2$
B	1	-1	1	-1	1	-1	1	-1	z	
E_1	1	ε	i	$-\varepsilon^*$	-1	$-\varepsilon$	$-i$	ε^*		
	1	ε^*	$-i$	$-\varepsilon$	-1	$-\varepsilon^*$	i	ε	$(x, y);$ (R_x, R_y)	
E_2	1	i	-1	$-i$	1	i	-1	$-i$		
	1	$-i$	-1	i	1	$-i$	-1	i		$(x^2 - y^2, xy)$
E_3	1	$-\varepsilon^*$	$-i$	ε	-1	ε^*	i	$-\varepsilon$		
	1	$-\varepsilon$	i	ε^*	-1	ε	$-i$	$-\varepsilon^*$		(xz, yz)

The Cubic Groups

T	E	$4C_3$	$4C_3^2$	$3C_2$		$\varepsilon = \exp(2\pi i/3)$
A	1	1	1	1		$x^2 + y^2 + z^2$
	1	ε	ε^*	1		$(2z^2 - x^2 - y^2,$
E						
	1	ε^*	ε	1		$x^2 - y^2)$
T	3	0	0	-1	$(R_x, R_y, R_z); (x, y, z)$	(xy, xz, yz)

T_d	E	$8C_3$	$3C_2$	$6S_4$	$6\sigma_d$		
A_1	1	1	1	1	1		$x^2 + y^2 + z^2$
A_2	1	1	1	-1	-1		
E	2	-1	2	0	0		$(2z^2 - x^2 - y^2, x^2 - y^2)$
T_1	3	0	-1	1	-1	(R_x, R_y, R_z)	
T_2	3	0	-1	-1	1	(x, y, z)	$(xy, xz, yz))$

T_h	E	$4C_3$	$4C_3^2$	$3C_2$	i	$4S_6$	$4S_6^5$	$3\sigma_h$	$\varepsilon = \exp(2\pi i/3)$
A_g	1	1	1	1	1	1	1	1	$x^2+y^2+z^2$
E_g	1	ε	ε^*	1	1	ε	ε^*	1	$(2z^2-x^2-y^2,\, x^2-y^2)$
	1	ε^*	ε	1	1	ε^*	ε	1	
T_g	3	0	0	-1	3	0	0	-1	(R_x, R_y, R_z) (xz, yz, xy)
A_u	1	1	1	1	-1	-1	-1	-1	
E_u	1	ε	ε^*	1	-1	$-\varepsilon$	$-\varepsilon^*$	-1	
	1	ε^*	ε	1	-1	$-\varepsilon^*$	$-\varepsilon$	-1	
T_u	3	0	0	-1	-3	0	0	1	(x, y, z)

O_h	E	$8C_3$	$6C_2$	$6C_4$	$3C_2(=C_4^2)$	i	$6S_4$	$8S_6$	$3\sigma_h$	$6\sigma_d$	
A_{1g}	1	1	1	1	1	1	1	1	1	1	$x^2+y^2+z^2$
A_{2g}	1	1	-1	-1	1	1	-1	1	1	-1	
E_g	2	-1	0	0	2	2	0	-1	2	0	$(2z^2-x^2-y^2,\, x^2-y^2)$
T_{1g}	3	0	-1	1	-1	3	1	0	-1	-1	(R_x, R_y, R_z)
T_{2g}	3	0	1	-1	-1	3	-1	0	-1	1	(xz, yz, xy)
A_{1u}	1	1	1	1	1	-1	-1	-1	-1	-1	
A_{2u}	1	1	-1	-1	1	-1	1	-1	-1	1	
E_u	2	-1	0	0	2	-2	0	1	-2	0	
T_{1u}	3	0	-1	1	-1	-3	-1	0	1	1	(x, y, z)
T_{2u}	3	0	1	-1	-1	-3	1	0	1	-1	

The blocked-off section represents the pure rotation O group.

The Icosahedral Group

I_h	E	$12C_5$	$12C_5^2$	$20C_3$	$15C_2$	i	$12S_{10}$	$12S_{10}^3$	$20S_6$	15σ		
A_g	1	1	1	1	1	1	1	1	1	1		$x^2+y^2+z^2$
T_{1g}	3	$\frac{1}{2}(1+\sqrt5)$	$\frac{1}{2}(1-\sqrt5)$	0	-1	3	$\frac{1}{2}(1-\sqrt5)$	$\frac{1}{2}(1+\sqrt5)$	0	-1	(R_x, R_y, R_z)	
T_{2g}	3	$\frac{1}{2}(1-\sqrt5)$	$\frac{1}{2}(1+\sqrt5)$	0	-1	3	$\frac{1}{2}(1+\sqrt5)$	$\frac{1}{2}(1-\sqrt5)$	0	-1		
G_g	4	-1	-1	1	0	4	-1	-1	1	0		(xz, yz, xy)
H_g	5	0	0	-1	1	5	0	0	-1	1		$(2z^2-x^2-y^2, x^2-y^2),$ (xy, yz, zx)
A_u	1	1	1	1	1	-1	-1	-1	-1	-1		
T_{1u}	3	$\frac{1}{2}(1+\sqrt5)$	$\frac{1}{2}(1-\sqrt5)$	0	-1	-3	$-\frac{1}{2}(1-\sqrt5)$	$-\frac{1}{2}(1+\sqrt5)$	0	1	(x, y, z)	
T_{2u}	3	$\frac{1}{2}(1-\sqrt5)$	$\frac{1}{2}(1+\sqrt5)$	0	-1	-3	$-\frac{1}{2}(1+\sqrt5)$	$-\frac{1}{2}(1-\sqrt5)$	0	1		
G_u	4	-1	-1	1	0	-4	1	1	-1	0		
H_u	5	0	0	-1	1	-5	0	0	1	-1		

The blocked-off section represents the pure rotation I group.

The Axial Groups

$C_{\infty v}$	E	$2C_\infty^\Phi$	$\cdots$	$\infty\sigma_v$		
$A_1 \equiv \Sigma^+$	1	1	$\cdots$	1	z	x^2+y^2, z^2
$A_2 \equiv \Sigma^-$	1	1	$\cdots$	-1	R_z	
$E_1 \equiv \Pi$	2	$2\cos\Phi$	$\cdots$	0	$(x,y); (R_x, R_y)$	(xz, yz)
$E_2 \equiv \Delta$	2	$2\cos 2\Phi$	$\cdots$	0		(x^2-y^2, xy)
$E_3 \equiv \Phi$	2	$2\cos 3\Phi$	$\cdots$	0		
$\cdots$		$\cdots$	$\cdots$	$\cdots$		

$D_{\infty h}$	E	$2C_\infty^\Phi$	$\cdots$	$\infty\sigma_v$	i	$2S_\infty^\Phi$	$\cdots$	∞C_2		
Σ_g^+	1	1	$\cdots$	1	1	1	$\cdots$	1		x^2+y^2, z^2
Σ_g^-	1	1	$\cdots$	-1	1	1	$\cdots$	-1	R_z	
Π_g	2	$2\cos\Phi$	$\cdots$	0	2	$-2\cos\Phi$	$\cdots$	0	(R_x, R_y)	(xz, yz)
Δ_g	2	$2\cos 2\Phi$	$\cdots$	0	2	$2\cos 2\Phi$	$\cdots$	0		(x^2-y^2, xy)
$\cdots$			$\cdots$				$\cdots$			
Σ_u^+	1	1	$\cdots$	1	-1	-1	$\cdots$	-1	z	
Σ_u^-	1	1	$\cdots$	-1	-1	-1	$\cdots$	1		
Π_u	2	$2\cos\Phi$	$\cdots$	0	-2	$2\cos\Phi$	$\cdots$	0	(x,y)	
Δ_u	2	$2\cos 2\Phi$	$\cdots$	0	-2	$-2\cos 2\Phi$	$\cdots$	0		
$\cdots$			$\cdots$				$\cdots$			

Capital Greek letters refer to molecular states and the subscripts g and u, respectively, refer to 'gerada' (even) and 'ungerada' (uneven), related to parity of the corresponding quantum states. The $\pm$ sign on Σ states indicates the character under the σ_v operation.

Appendix B

Irreps for the O_h and D_{4h} Groups, their Dimensions and Notations[1]

O_h Group

Chemical	E	Bethe	BSW
$A_{1g,u}$	1	$\Gamma_1^\pm$	Γ_1, Γ_1'
$A_{2g,u}$	1	$\Gamma_2^\pm$	Γ_2, Γ_2'
$E_{g,u}$	2	$\Gamma_3^\pm$	$\Gamma_{12}, \Gamma_{12}'$
$T_{1g,u}$	3	$\Gamma_4^\pm$	$\Gamma_{15}', \Gamma_{15}$
$T_{2g,u}$	3	$\Gamma_5^\pm$	$\Gamma_{25}', \Gamma_{25}$
$E_{1/2}^\pm$	2	$\Gamma_6^\pm$	
$E_{5/2}^\pm$	2	$\Gamma_7^\pm$	
$G^\pm$	4	$\Gamma_8^\pm$	

D_{4h} Group

$A_{1g,u}$	1	$\Gamma_1^\pm$	M_1, M_1'
$B_{1g,u}$	1	$\Gamma_3^\pm$	M_2, M_2'
$B_{2g,u}$	1	$\Gamma_4^\pm$	M_4, M_4'
$E_{g,u}$	2	$\Gamma_5^\pm$	M_5', M_5'

---

[1]In these listings the (u, g) subscripts correspond to $(+, -)$ superscripts, respectively. The (prime, unprime) markings likewise correspond to $(+, -)$ superscripts, respectively. E-column values of the character table (i.e., degeneracy) of each irrep are shown as well.

Bibliography
of Works in Group Theory
and Allied Topics

1. S.L. Altmann, *Band Theory of Solids: Introduction from the Point of View of Symmetry*, Oxford (1991).

2. E. Artin, *Modern Higher Algebra (Galois Theory)*, Compiled by A.A. Blank, NYU (1947)

3. G. Birkhoff and S. MacLane, *A Survey of Modern Algebra*, 5th ed., A.K. Peters, (1997).

4. D.M. Bishop, *Group Theory and Chemistry*, Dover (1973).

5. C.J. Bradley and A.P. Cracknell, *The Mathematical Theory of Symmetry in Solids*, Oxford (1972).

6. D.M. Burton, *Elementary Number Theory*, 3rd ed., W.C. Brown, (1994).

7. F.J. Budden, *The Fascination of Groups*, Cambridge (1972).

8. G. Burns, *Introduction to Group Theory with Applications*, Academic Press, (1977).

9. G. Burns and A.M. Glazer, *Space Groups for Solid State Scientists*, 2nd ed., Academic Press (1990).

10. W. Burnside, *Theory of Groups of Finite Order*, 2nd ed., Dover (1955).

11. J.F. Cornwell, *Group Theory in Physics*, Academic Press (1984).

12. F.A. Cotton, *Chemical Applications of Group Theory*, 3rd ed., Wiley Interscience (1993).

13. D.S. Dummit and R.M. Foote, *Abstract Algebra*, Prentice Hall (1991).

14. W.E. Deskins, *Abstract Algebra*, Dover, (1995).

15. B.N. Figgis, *Introduction to Ligand Fields*, Interscience (1966).

16. D.J.H. Garling, *A Course in Galois Theory*, Cambridge (1986).

17. W. Griener and B. Muller, *Quantum Mechanics, Symmetries*, Springer (1989).

18. J.S. Griffith, *Theory of Transition Metal Ions*, Cambridge (1961).

19. M. Hamermesh, *Group Theory*, Addison-Wesley (1991).

20. G.G. Hall, *Applied Group Theory*, American Elsevier (1967).

21. L.H. Hall, *Group Theory and Symmetry in Chemistry*, McGraw-Hill (1969).

22. V. Heine, *Group Theory and Quantum Mechanics*, Dover, (1993).

23. N.F.M. Henry and K. Lonsdale, eds. *International Tables for X-ray Crystallography*, Kynoch, (1952).

24. R.M. Hochstrasser, *Molecular Aspects of Symmetry*, W. A. Benjamin (1966).

25. N.F.M. Henry and K. Lonsdale, eds., *International Tables for X-ray Crystallography*, Kynoch, London (1969)

26. I.N. Herstein, *Abstract Algebra*, 3rd ed., Prentice Hall (1996).

27. T. Inui, Y. Tanabe and Y. Onodera, *Group Theory and Applications in Physics*, Springer (1995).

28. I. Jansen and M. Boon, *Theory of Finite Groups, Application in Physics*, North-Holland (1967).

29. T. Janssen, *Crystallographic Groups. Application in Physics*, North-Holland (1973).

30. H. Jones, *Theory of Brillouin Zones and Electronic States in Crystals*, North-Holland (1960).

31. A.W. Joshi, *Elements of Group Theory for Physicists*, Wiley-Eastern (1983).

32. R.S. Knox and A. Gold, *Symmetry in the Solid State*, W.A. Benjamin (1964).

33. G.F. Koster, *Space Groups and Their Representations*, Academic Press (1957).

34. J.S. Lamont, *Applications of Finite Groups*, Dover (1989).

35. L.D. Landau, E.M. Lifshitz and L.P. Pitaevskii, *Quantum Mechanics, Non-Relativistic Theory*, Pergamon (1977).

36. J. Landin, *An Introduction to Algebraic Structures*, Dover (1989).

37. W.J. LeVeque, *Fundamentals of Number Theory*, Dover (1977).

38. L. Mariot, *Group Theory and Solid State Physics*, Prentice Hall (1962).

39. P.H.E. Meijer and E. Bauer, *Group Theory*, North-Holland (1962).

40. P.J. McCarthy, *Algebraic Extensions of Fields*, Blaisdell (1966).

41. R. McWeeny, *Symmetry and Introduction to Group Theory*, MacMillan (1963).

42. N.D. Mermin, "The Space Groups of Icosahedral Quasicrystals and ...," *Reviews of Modern Physics*, **64**, 3–20 (1992).

43. F.D. Murnagham, *The Theory of Group Representations*, Dover (1963).

44. P.M. Neumann, G.A. Stoy and E.C. Thompson, *Groups and Geometry*, Oxford (1994).

45. A.S. Nowick, *Crystal Properties vis Group Theory*, Cambridge (1995).

46. A. Nussbaum, *Applied Group Theory for Chemists, Physicists and Engineers*, Prentice Hall (1971).

47. J.F. Nye, *Physical Properties of Crystals*, Oxford (1985).

48. L.S. Pontryagin, *Topological Groups*, Gordon and Breach (1966).

49. D.J.S. Robinson, *A Course in the Theory of Groups*, 2nd. ed., Springer (1995).

50. J.J. Rotman, *Introduction to the Theory of Groups*, 4th ed., Springer (1995).

51. W.R. Scott, *Group Theory*, Dover, (1987).

52. J. Slater, *Symmetry and Energy Bands in Crystals*, Dover (1972).

53. M. Tinkham, *Group Theory and Quantum Mechanics*, McGraw-Hill (1964).

54. Wu-Ki Tung, *Group Theory in Physics*, World Scientific (1985).

55. B.L. van der Waerden, *Group Theory and Quantum Mechanics*, Springer (1974).

56. S. Warner, *Modern Algebra*, Dover (1990).

57. M. Weissbluth, *Atoms and Molecules*, Academic Press (1978).

58. B.S. Wherret, *Group Theory for Atoms, Molecules and Solids*, Prentice Hall (1986).

59. E.P. Wigner, *Group Theory*, Academic Press (1959).

60. H. Weyl, *The Theory of Groups and Quantum Mechanics*, Dover (1931).

61. H. Weyl, *Symmetry*, Princeton University Press (1952).

62. P.Y. Yu and M. Cardona, *Fundamentals of Semiconductors*, Springer (1996).

Index